CCRP
Publication
Series

This is a continuation in the series of publications produced by the Center for Advanced Concepts and Technology (ACT), which was created as a "skunk works" with funding provided by the CCRP under the auspices of the Assistant Secretary of Defense (C3I). This program has demonstrated the importance of having a research program focused on the national security implications of the Information Age. It develops the theoretical foundations to provide DoD with information superiority and highlights the importance of active outreach and dissemination initiatives designed to acquaint senior military personnel and civilians with these emerging issues. The CCRP Publication Series is a key element of this effort.

Check our website for the latest CCRP activities and publications.

www.dodccrp.org

DoD C4ISR Cooperative Research Program

Senior Civilian Official and Chief Information Officer, OASD (C3I)
      Mr. Arthur L. Money
Director, Performance Assessment
      Dr. Margaret E. Myers
Director, Research and Executive Agent for CCRP
      Dr. David S. Alberts

Library of Congress Cataloging-in-Publication Data

Howard, Nigel.
  Confrontation analysis : how to win operations other than war /
Nigel Howard.
    p. cm. — (CCRP publication series)
    Includes bibliographical references.
    ISBN 1-893723-00-3 pbk
    1. United States—Armed Forces—Operations other than war.
  I. Title. II. Series.
  UA23.H568   1999
  355.4—dc21   99-22842
  August 1999   CIP

CONFRONTATION ANALYSIS:

HOW TO WIN
OPERATIONS OTHER THAN WAR

NIGEL HOWARD

TABLE OF CONTENTS

LIST OF TABLES AND FIGURES

v

Acknowledgments

This book is based on research commissioned by the UK Defence Evaluation and Research Agency (DERA) at the beginning of 1997 into how Confrontation Analysis might be applied to Peace Support Operations. The idea of turning the research into a book came from Dr. David S. Alberts, Director of Research and Executive Agent for the C4ISR Cooperative Research Program. For this I am most grateful to him.

Andrew Tait (now with SAIC, then with DERA) was responsible for suggesting the original research project. His suggestion was taken up by George Brander and Paul Willis of DERA, while Peter Murray-Jones also took a keen interest. Many persons with knowledge or experience of peace support managed to find time to assist with the research, including MajGen Pennefeather (Commandant-General of the Royal Marines), Brig Alastair Duncan, Col Robert Stewart (Retd), LtCol Philip Wilkinson, Brig B.C. Lambe, Prof K.C. Bowen, Michael Codner and Edward Foster (Royal United Services Institute for Defence Studies), Prof James Gow, Indjana Harper and Vesna Domani-Hardy. Capt David Fifield helped to make military contacts, and Brig R. Lambe (Retd), Steve Lea, George Rose and Graham Mathieson were among many at DERA who helped by responding to the ideas. The basic ideas have themselves been worked out in collaboration with Peter Bennett (now at the UK Department of Health), Prof Jim Bryant and

Prof Morris Bradley, as well as through interactions with numerous academic and business colleagues over many years.

Finally, I must express my gratitude to Dr. Richard E. Hayes and Richard Layton, Evidence Based Research, Inc., and their staff, Lynne Jennrich and Margita Rushing for their work on drafts of the book, and Meg Rittler for her work on the figures and cover design.

PREFACE

This book presents a simple idea. A Peace Operations campaign (or Operation Other Than War) should be seen as a *linked sequence of confrontations*—in contrast to a traditional, warfighting campaign, which is a *linked sequence of battles.* The objective in each confrontation is to bring about certain "compliant" behavior on the part of other parties, until in the end the campaign objective is reached. This is a state of sufficient compliance to enable the military to leave the theater.

If this simple idea is accepted, we can show how the new technique of Confrontation Analysis (derived from Game Theory via a development called Drama Theory) can be applied. Thus we can show how to *win* an Operation Other Than War.

Since this book was written, further research carried out in the Bosnia theater has clearly revealed that SFOR commanders—from platoon commanders to the overall theater commander—are doing it already. They are winning confrontations, or campaigns, made up of linked sequences of confrontations on a day-to-day basis.

They are doing it, however, without a clear, uniform system of concepts specifically designed for a confrontational campaign. Using practical good sense, they are instead taking doctrinal concepts developed primarily for warfighting, and adapting them for use in confrontations. For example, they are using concepts of artillery targeting to plan how to "target" noncompliant parties (i.e., a local Mayor and police chief who are refusing to provide security for returning refugees from a different ethnic group).

Such common-sensical adaptation of standard warfighting systems and concepts is admirable. And it works.

We believe, however, that a system that treats confrontations as confrontations, distinguishing them from battles both conceptually and in terms of planning procedures, will enable striking improvements to be made so that our forces become still more effective. In particular, it will make it possible to use the powerful techniques of Confrontation Analysis described in this book.

These are ideas that still have to win acceptance. This book aims to lay them out for your consideration.

CHAPTER 1

THE NEED TO RECONCEPTUALIZE OPERATIONS OTHER THAN WAR

"War is a mere continuation of politics by other means."
—Clausewitz

From 1945 until 1989, U.S. and Allied defense forces took as their first priority preparations for a superpower conflict that never actually occurred. Thus billions of dollars were spent in pursuit of a theoretical construct.

Since 1989, the political assumptions on which this construct was based have shifted. High-intensity superpower conflict, the first and most important contingency we had to prepare for, vanished from the immediate agenda because only one superpower remained. The result was that instead of a conflict conceived and planned for using theory alone, defense forces now must prioritize lesser threats of which they have real experience.

Although this is real experience, as distinct from a theoretical deduction, it is hard to make sense of in traditional military terms. To the theorist using

1

traditional models, it is messy and disappointing. This has put the defense community in a difficult position; it has to ask for and use public money in pursuit of a vision, "persuasive for peace, decisive in war, preeminent in any form of conflict" (Joint Chiefs of Staff, 1997), whose conceptual and doctrinal underpinnings it needs to clarify.

What is this messy, real-world experience?

Traditional Clausewitzian missions do come up, as in the case of the Gulf War (*Operation Desert Storm*). But the traditional character of even this mission seems, in retrospect, less than obvious. Traditional military objectives were, in the event, attained easily, with minimal loss of life to the Allies and frightful losses to Iraq. Yet the political problem was poorly resolved.

Other missions are not at all Clausewitzian. Peace operations in general (see Alberts and Hayes [1995], Maxwell [1997], and Wentz [1997]) have to do with political stability and humanitarian assistance, rather than physically compelling an enemy to submit to our will. Objectives in these missions change and develop as the mission goes on. Cooperation is required between military forces and non-governmental organizations that share responsibility for mission objectives and must take them over when the military leaves. Economic, political, and psychological pressures are important. Military commanders must negotiate with conflicting parties rather than fight them.

Despite these non-Clausewitzian characteristics, one point made by Clausewitz (1968, 1st edition 1832) is more relevant than ever. War and war preparations are political acts, whose nature varies with their political background and aim.

Thus, to understand war in the era beginning after the Cold War, and to begin to provide better conceptual underpinning for defense budgets and plans, we will try to understand the new political structures that characterize this era.

THE NEW WORLD ORDER

What are defense forces now defending?

To be frank, what happened after 1989 was the economic and military triumph of a single, unified world system that U.S. President George Bush called the New World Order (Bush, 1990). This system makes surprisingly specific and detailed demands:

> *...that all must be capitalist, democratic, tolerant, non-racist, and non-sectarian; allow equal opportunities and freedom of speech and the press; protect human rights; not let governments greatly incommode world trade or capital movements; and be peaceful except when enforcing these demands.*

We admit, of course, that no sooner had Bush uttered the words "New World Order" than the concept was denounced as an absurd chimera by every commentator. What is significant is that all recognized at once what he meant and were able, unanimously and without coordination, to agree as to what it was they considered absurd. Their outraged denials, repeated at intervals ever since, merely gave greater substance and wider dissemination to the concept, helping to make it the powerful coordinating framework it has become.

The coordinating power of this universal understanding is immense. When the United States and its allies seek to impose peace and order anywhere in the world, all know more or less what kind of order they are demanding. Moreover, there is a definite tendency for populations everywhere to demand it for themselves, despite local efforts to persuade them not to.

Russia or China may still provide centers of resistance to the New World Order. Their ability to do so is doubtful, and decreases as they become more dependent upon and involved in the system. As they and others join it, the system will of course evolve. It is essentially dynamic; however, fundamental change can be expected to be continuous and evolutionary in the manner of capitalist development, rather than violent and disruptive.

The system now called the New World Order can be seen, in retrospect, to have been growing within the shell of the old, bipolar world since 1945. The old Clausewitzian assumptions that war takes place between more-or-less equal nations is carried out by military professionals leading citizen soldiers, and is largely military in character as distinct from political, held true exceptionally, if at all, from 1945 until 1989. This is because, under the guise of the Free World combating Communism, Bush's New World Order was developing.

Now that the system has emerged into the open, how does it increasingly enforce its disciplines?

Minor, particularly economic transgressions are punished by economic sanctions. Transgressions that are not only violent but also sufficiently gross, well-publicized, or economically damaging, such as in the Gulf, Bosnia, Ireland, or Oklahoma, evoke forceful responses. Defense forces are used then.

Among forceful interventions, the Gulf War generally was counted a success, Somalia a failure, and Bosnia a mixture of failure followed by the hope of eventual success after the United States came in and got tough. Northern Ireland has taken a turn for the better. The threat of internal terrorism (e.g., the Oklahoma bomb, the Tokyo subway gas attack) resembles an ongoing war against an ever-varying enemy, with new methods of attack continually met by new methods of defense. The Arab-Israeli problem seems incapable of resolution as long as the United States, because of internal political divisions, remains unable to intervene with sufficient force.

How Can the Strong Defeat the Weak?

Even over the Gulf War, counted a success, questions are asked. At the end, why did Allied forces refrain from unseating Saddam Hussein? Rogers (1997) suggests that it was because the Allies feared Saddam's biological weapons, which he had delegated to the control of his commanders in the event that overall central command was lost. Although we do not believe this, it shows the kind of power that might potentially be wielded by a player that has been comprehensively defeated.

Somalia and Bosnia expose in other ways the inadequacy of conventional military responses to new kinds of threat. Internal terrorist threats show this more clearly. Consider not only the Oklahoma and World

Trade Center bombs, the IRA's economic targeting, and the Tokyo subway attack, but the highly successful destruction of the Bank of Sri Lanka by Tamil Tigers. In each case, modern technology and terrorist or guerrilla tactics have enabled the weak to take on the strong.

The problem of defense in the modern world is the paradoxical one of finding ways for the strong to defeat the weak. Obviously, this should not be an impossible task; however, doing it with maximum effectiveness and minimal loss of life and treasure does require a new approach.

Confrontations, Not Battles

What kind of success can the weak aim for against the strong?

The weak can never hope to defeat the New World Order by fighting battles, the predominant mode of warfare that defense forces prepared for for 40 years. At most, the weak can hope for single, isolated victories against local outposts. Some may see these as significant because they believe the system is so unpopular that their attacks will be copied and so fragile it will then collapse. Others aim, more realistically, to get away with something the system forbids: to commit genocide against local ethnic neighbors (Rwanda, Bosnia); to annex or transform a nation without a democratic mandate to do so (Iraq, the IRA before Easter 1998); or to grab weapons to loot and rape neighbors (urban riots, Albania).

In opposing such rebels, the system's guardians must make or reject demands. They must say either: "You must give up..." (genocide, aggression, etc.)

or "We refuse to…" (publish your manifesto, release your comrades from prison, etc.). In either case, defense forces must sustain and win not a battle but what we may call a confrontation. This is a situation in which victory consists of compelling, persuading, or inducing others to submit to our will without, if possible, using violence, although violence, as well as other threats and inducements, must be a credible part of our armory.

Forty years of preparation for the world's greatest battle have not equipped defense forces for winning confrontations. Our aim in this book is to begin to remedy this comparative lack of preparation.

We propose to investigate confrontations in general using the technique of confrontation analysis, by which the commander of a peace operation may plan and execute a strategy for fighting and winning it. We would like to claim, in fact, that by using this technique it is possible to take a more logical, ordered, and defensible approach to this problem than to the analogous problem of winning battles.

This is not because emotion, friction, irrationality, and the Clausewitzian "fog of war" are less present in confrontations than they are in battles. On the contrary, they are found in all types of conflict. What we claim is that we have a method for analyzing these and other factors systematically and scientifically.

A GENERAL MODEL

Confrontation analysis does not apply a general model indiscriminately to every confrontation. On the contrary, we model each confrontation separately, picking out and taking advantage of its special features.

However, to show what a confrontation is, table 1 sets out a simple, general model of one taking place between the Allies and certain unidentified Rebels against the New World Order. Here, the Rebel position is that they should not be required to give in, but Allied forces should concede their demands, the precise nature of which, in this particular case, we do not specify. The Allied position is that the Rebels should give in without any concessions being made.

These positions are displayed in table 1 using the metaphor of a card-table. This is simple. It works as follows:

- Each player (participating party) holds certain cards, representing its yes/no policy options. In table 1 players are listed on the left, with their cards listed below their names (e.g., the Rebels hold the cards "Give in" and "Retaliate"; Allied forces hold the cards "Concede" and "Crush Rebels").

- In general, a player can play any combination of its cards (although some combinations may be infeasible). In this manner a player chooses and implements a policy.

After each player chooses a combination of its cards to play, a column of cards laid out on the card-table represents a projection of the situation as it would be

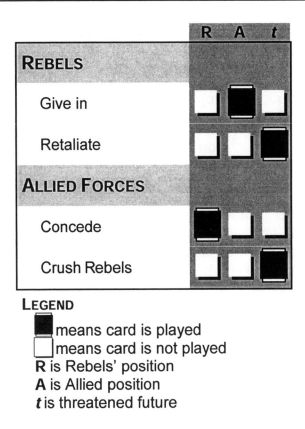

LEGEND
■ means card is played
□ means card is not played
R is Rebels' position
A is Allied position
t is threatened future

Table 1. Confrontation between Allied Forces and Rebels.

determined by those policy choices. The choice by a player to play a particular card is shown in table 1 by a heavy, framed cell representing a card; the choice not to play it is shown by a white cell. Column R, for example, represents the future in which the Allies concede the Rebels' demands and do not crush them. The Rebels, accordingly, do not give in (their demands having been met) and do not retaliate either (action to crush them not having been taken).

How Positions and Fallback Positions Are Shown in a Card-Table

This, of course, is precisely the Rebels' position (i.e., the solution the Rebels demand). The Rebels' position is shown by an appropriately labeled column.

The Allies' position (i.e., the Allied solution, that the Rebels give in and not retaliate, while the Allies neither concede nor crush them) is shown in column A.

Column *t* shows the threatened future: What the parties implicitly or explicitly threaten to do if their positions are not accepted. The Allies say, "If you don't give in, we'll crush you." The Rebels say, "If you try, we'll retaliate."

This concept addresses an item of utmost importance in a confrontation: what, in the last resort, a party conveys to all and sundry that it will do if its demands on the situation are not met. We call it the party's fallback position.

Note that the players' ways of conveying their fallback position, which is a particular kind of conditional intention, may be direct or indirect. A player may sometimes convey its fallback position by denying it. Imagine, for example, what it would mean if the Russian president declared he has no intention of invading a certain country, but that it should nevertheless discontinue its anti-Russian policies.

The implicit, although explicitly denied threat in such a statement would not necessarily be credible. It would, however, be clearly made. The important point is that at a climactic "moment of truth" in a confrontation, each party's fallback position (whether or not it is credible) is clearly conveyed to the others.

In a card-table model, parties' fallback positions are represented by certain selections from their own cards. Putting these selections together, we obtain a future, called the threatened future. This is column *t* in table 1. Here the Allied forces are implicitly threatening to crush the Rebels (often merely bringing armed forces into a theater is to implicitly threaten to use them); the Rebels are threatening retaliation in that case.

What This Model Represents

Many examples roughly fit our generalized model. In a (highly simplified) model of the Northern Ireland confrontation toward the end of the IRA's period of armed struggle, the Rebels could be the IRA. The Allied forces would be or the governments of Britain and Ireland. "Giving in" might then mean accepting the Anglo-Irish offer of negotiation in return for decommissioning arms. "Retaliation" might mean a continuing terrorist campaign. "Conceding" might mean handing over power to Sinn Fein and the IRA. "Crushing the Rebels" might mean continuing to fight terrorism.

This is a high-level example involving governments and national movements as players. But the same pattern appears at other levels. The first British troops in Bosnia, for example, took on the task of clearing roadblocks set up by different factions. We could fit our model to the kind of situations they faced. The Allied forces of our model would be the local force led by a British colonel. Rebels would be the local ethnic militia manning a particular roadblock. "Giving in" would mean dismantling the roadblock. "Conceding" would mean accepting its continuance. "Crushing the rebels" might have meant using tanks to eliminate the roadblock. "Retaliation" might have meant opening fire on British troops.

DO WE KNOW HOW TO DO CONFRONTATIONS?

How should a commander deal with a confrontation? Does he know how to win it? What does "winning" mean?

Although individual commanders have been successful in individual cases, there is no general, trained military competence in this area. Yet our argument is that winning confrontations is the clue to conducting and winning peace operations. Just as a traditional military campaign may be conceptualized as a linked sequence of battles, so a peace operation may, we suggest, be seen as a linked sequence of confrontations. Continuing the comparison: winning a traditional campaign consists of achieving overall mission objectives through a sequence of battles, even though objectives may not be fully met in each particular battle, and some battles may be avoided rather than fought. Winning a peace operation may be defined similarly. We merely have to replace the word "battle" with the word "confrontation." The aim in a peace operation is to achieve overall objectives through conducting, and as far as possible winning, a sequence of confrontations.

Yet compare the differences in preparedness between fighting a traditional campaign and a peace operation. Current military doctrine and training give excellent guidance as to how to conduct individual battles and plan a victorious campaign through a whole sequence of such battles. Complete, detailed instructions are not, of course, given. Training and doctrine provide guidelines, adaptable by the commander to meet changing circumstances. In this way a commander is taught how to plan a strategy and how to implement it by devolving responsibility for its various elements horizontally and vertically throughout his command. By comparison, with the guidelines available for many other human activities (e.g., setting up and running a government department) these guidelines must be judged to be highly scientific and effective.

Where do we find comparable guidelines on how to fight and win individual confrontations and whole sequences of them? Where is the training not only in how to devise and follow a strategy, but in how to devolve it to responsible units?

Figure 1 suggests an explanation for this comparative lack of trained understanding. It depicts the traditional relationship, implicitly assumed by Clausewitz and other theorists, between politicians and the military.

Traditionally it was assumed to be the job of politicians to resolve conflicts peaceably, if they could and wished to do so (see top part of figure 1). When conflict resolution failed or it was decided to use force instead, the politicians directed the military accordingly, instructing it to forcibly achieve political objectives (see lower part of figure 1). The job for the military was to conduct armed conflict and report back to politicians on their progress. This and only this was what the military was trained to do; actual, armed conflict defined military professional competence.

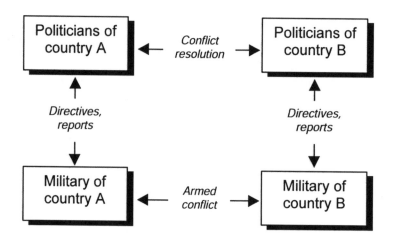

Figure 1. Traditional relationship between politicians and military.

The development of nuclear weapons following World War II caused one kind of change in this model: the chain of command was modified to ensure that politicians rather than generals made the decision, if any, to escalate to nuclear war. That decision was felt to be, as Clemenceau reportedly said of war in general, too serious a business to be left to generals. Thus, politicians began to intervene in decisions previously left to the military. Such intervention became more frequent as improved communications made it more feasible. It was extended to more types of conflict as the feeling grew that any use of force by a superpower or ally of a superpower might start a process of escalation; theoretical mechanisms accounting for such escalation were spelt out by Kahn (1965). In this way, nuclear weapons gave rise to a doctrinal concept of limited war.

At the same time, another kind of change was taking place. Members of nuclear alliances no longer could be challenged by forces strong enough to do so, because of fear of nuclear escalation. The only challenge that could still be mounted against them was a challenge by the weak, such as the challenge of low intensity conflict faced by Napoleon's armies in Spain.

As we have said, meeting a challenge mounted by the weak is essentially a matter of conducting a sequence of confrontations. Objectives cannot be achieved by force alone: the weak are not strong enough and the strong face an enemy that avoids decisive encounters by vanishing into the environment. Although such operations are conducted by the military, they can succeed only by reaching a political solution, and this solution cannot be reached by high-level decision makers alone because it must be grounded in the hopes and fears of the foot-soldiers fighting on behalf of the weak.

Consequently, the move to this kind of conflict had an opposite result to the move to limited war; it meant that the military was required to take over many of the conflict resolution activities traditionally assigned to politicians. Low-intensity conflict, it turned out, was too messy, detailed, local, and political to be left entirely to politicians to resolve.

Again, low-intensity conflict usually poses no direct threat to the integrity of the powerful nations involved, but only an indirect threat through the undermining of international order. Because of this, the objectives of low-intensity conflict are often far from clear. Often objectives are ill-defined resultants of compromise at the political level between different groups or different nations in a coalition. As a result, the political directions handed to the military are vague and ambiguous.

While limited war and low-intensity conflicts comprised most of the reality faced by defense forces in the post-war era, it was imperative for them, while the Cold War lasted, to train and study for the high-intensity, total superpower conflict that never actually occurred. This was their most serious task, on which national survival and the future of the world ultimately depended. The fact that this war never happened may, paradoxically, be a result of the fact that it was so soundly prepared for. This cannot be accounted a failure. It was a success.

The price of this success was a comparative lack of doctrine and training for limited war and low-intensity conflict. This is the lack we hope to begin, at least, to fill.

How To Do It, In Outline

Confrontation analysis, as said, proceeds by building specific models incorporating the specific details of each confrontation. It can be used to build models at each level, strategic, operational, and tactical. These models can be linked to enable a commander's strategy to be devolved into strategies for each subordinate level and linked to the strategies of players related horizontally to his command, such as coalition partners or non-governmental organizations.

Models so constructed will be strategic, looking forward to winning a whole operation through a sequence of confrontations. They will be readily capable of changing to meet changing contingencies, and such changes will be readily propagated through the system of linked models.

By presenting such a system in a simple, clear way, accessible to all professionals involved, we would hope eventually to provide doctrinal guidance and training for the at-present, vaguely defined task of peace support that is at least as good as that provided for fighting battles.

We would propose that eventually a theater commander in a peace support operation would have analysts on his staff trained in confrontation analysis. At each stage, starting with his first notification of the mission he is tasked with, these specialists would model the problem at each command level.

The commander's first need is, in general, to understand the problem facing his political masters, because understanding the intent of his commander is necessary for him to understand his own mission. To help him, analysts would model the peace mission as part of the world-political problem and the specific peace-support problem, as he sees it.

When a commander receives a specific mission, perhaps in vague, nonspecific terms that result from compromises between political actors, analysts would work with him to model the specific confrontations he is directly responsible for handling. Such analysis, incorporating the commander's own assumptions, should yield the two following results:

- Establishment of a sequence of steps to apply requisite pressure on other parties to bring them into compliance with the commander's position (i.e., mission objectives). We shall see how to derive this sequence based on dilemmas (change-inducing points of stress) we and others face, and how these are handled to achieve our objectives.

- Invention of an immersive role-playing exercise through which the commander and other officers involved can digest and criticize the analysis, rehearse their interactions with other parties, and become familiar with other-party points of view.

Analysts would then work with lower-level commanders and representatives of other components to build devolved and horizontally linked models. Both the analysis and the role-playing based on it would be updated as the confrontation evolves. The role-playing would be a kind of war-gaming, but with a firm analytic basis.

This, in outline, is the system we hope eventually will fill the present gap in doctrine and training for Operations Other Than War (OOTW). Chapter 6 will show how it might work. It contains the Frontline Play, an attempt to dramatize the situation of an imaginary commander of a peace support operation in a fictitious country who decides to analyze his problem using confrontation analysis.

Meanwhile, chapters 2 and 3 give a detailed account of what confrontation analysis means.

SUMMARY OF CHAPTER 1

Following the end of the Cold War, defense priorities generally have shifted to a new kind of mission, broadly described as defending the New World Order against Rebels who are militarily weaker. To understand this kind of mission, the concept of winning a campaign by fighting a linked sequence of battles needs to be replaced with that of winning an OOTW by conducting a linked sequence of confrontations.

A simple, card-table model of a general confrontation shows the position and fallback position of each party. A commander's mission is to get all parties' willing compliance with his position. In doing this, he needs to apply pressure on other parties in ways traditionally thought of as being a politician's job. A professional approach to this requirement is needed. It can be supplied by confrontation analysis.

CHAPTER 2

HANDLING A MULTILEVEL CONFRONTATION: THE SIX-PHASE MODEL

Winning an Operation Other Than War (OOTW), we have said, is a matter of conducting and winning a linked sequence of confrontations. The question is: How does a commander form and implement a strategy for this?

This is the key question. To begin to prepare an answer, in this chapter we will look at the process a confrontation goes through, leading it to being resolved or breaking out into conflict. The commander's aim is, of course, to have it resolved on his terms.

His strategy for achieving this is not primarily a matter of physical activities, as it is with warfighting. It is a matter of communication. At the same time, his possession of a credible capacity for warfighting is usually central to it.

As with warfighting, his strategy needs to be implemented on many levels. He should be able to give directives that link together the many confrontations occurring on different levels and implement them through a cohesive, multilevel strategy. To make this point more clearly, we begin with a concrete, albeit simplified, example.

EXAMPLE: REMOVING ROADBLOCKS

Table 2 shows a card-table model of the situation faced by a UN commander who has been tasked to make a certain ethnic militia force stick to an agreement it has signed. This model differs from the prototype in chapter 1 (see table 1) in that both parties are taking the same position. This does not, however, mean they have no further problems. The numbers alongside players' names in this model represent their preference rankings (the order in which they prefer the three futures shown). The most preferred future for that player is given the number 1, the next most preferred, number 2, and so on.

We are making the following assumptions:

> **Assumption 1**—The agreement states that the militia should cease ethnic cleansing (i.e., attacking villages and relocating members of another ethnic community) and allow free movement of people. The commander's cards include the following options: (a) give or withhold support and aid to the militia, which constitutes the effective local government; (b) if the militia disallows free movement, the commander may force it by removing roadblocks and escorting travelers; (c) if the militia flagrantly violates the agreement, the commander may bombard the militia's headquarters, as he was doing until the militia signed the agreement.

Because the militia has now signed the agreement, it shares the commander's position that the agreement be kept (column P); however, the commander suspects that the militia does not actually intend to

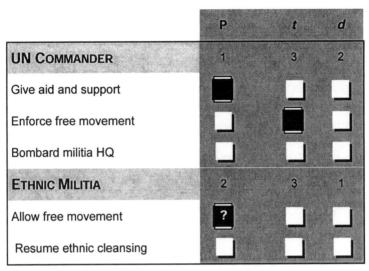

	P	*t*	*d*
UN COMMANDER	1	3	2
Give aid and support	■	□	□
Enforce free movement	□	■	□
Bombard militia HQ	□	□	□
ETHNIC MILITIA	2	3	1
Allow free movement	?	□	□
Resume ethnic cleansing	□	□	□

LEGEND

■ means card is played
□ means card is not played
P is joint position of UN commander and militia
t is threatened future
d is default future
? means playing this card is not preferred
# indicates preference ranking (1 is most preferred)

Table 2. UN commander confronts ethnic militia.

allow free movement, which would encourage displaced villagers to return to their homes, explaining the question mark placed on this card. Hence, the confrontation is not really resolved. [Note: The word "intend" needs clarifying here. In what sense does a complex, multifarious organization such as an ethnic militia intend anything? Answer: A player's objectives and beliefs are considered to be the result of internal confrontations between internal factions (subplayers belonging to and identifying with the player). In our case, the militia leadership (one subplayer) may want to allow free movement, but the leadership's followers, who are local villagers operating locally (and constitute another subplayer), do not. The commander, we assume, suspects that the followers' view may prevail. In this sense, the commander suspects the militia intends to defect from the agreement.]

> **Assumption 2**—The current threatened future (column t) is not the threatened future whose actual implementation (bombarding of the militia headquarters) led to the signing of the agreement. It is the future the commander is now threatening to enter on (by implementing his part of it) if the militia does not allow free movement. Simultaneously the militia is threatening to implement its part if the commander does not give aid and support.

The term "default future," attached to column d, is a general term for the future now being implemented through the parties' current policies, and which will therefore continue unless and until those policies are changed. The militia has given up ethnic cleansing, but it has not yet allowed free movement. The commander has not started giving aid and support; nor has he started forcibly removing roadblocks.

Note that the cards "bombard militia HQ" and "resume ethnic cleansing" are included in the model although they are not played in any character's position or fallback position, nor in the default future. They are there because they are active possibilities in players' minds. A card-table model should include all cards the players consider relevant. It must include all cards needed to define players' positions and the threatened future; but it may include more, as in this case.

In summary, the UN commander has bombarded the ethnic militia until it agreed to give up ethnic cleansing and allow free movement. The commander then promised to give the ethnic militia aid and support. He has not done so yet, and the militia has not yet allowed free movement; thus, the agreement has not been fully implemented.

Linkages Between Different Levels

Next, to illustrate linkages between two levels of command, table 3 models the situation of a local battalion commander faced with a roadblock. The local militia's position (M) is that the roadblock stays. In light

	M	C	t	t^l	d
LOCAL COMMANDER	3	1	2	4	3
Forcibly remove roadblock	□	□	■	■	□
Call in air strikes	□	□	□	■	□
LOCAL MILITIA	1	3	2	4	1
Remove roadblock	□	■	□	□	□
Fire on UN troops	□	□	□	■	□

LEGEND
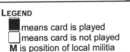
■ means card is played
□ means card is not played
M is position of local militia
C is local commander's position
t is threatened future
t^l is a possible alternative threatened future
d is default future
# indicates preference ranking (1 is most preferred)

Table 3. Local commander confronts roadblock.

of his commander's intent (as shown in table 2), the local commander's position is that it be removed or he will forcibly remove it (t). If the militia then fires on his troops from surrounding hilltops, the commander will call in air strikes (t').

Note that the players are considering two possible threatened futures. This is because the ethnic militia has not made its fallback position clear (i.e., whether it will fire on UN troops if roadblocks are forcibly removed). This ambiguity indicates that the confrontation is still in the process of building up to a climax. At a proper climax, all positions are made clear.

The local commander's card-table is linked to his commander's card-table; it represents the implementation at the local level of the commander's overall position and fallback position.

This statement is clearly true in a general way; however, it is important that the exact manner of the linkage be decided by the local commander based on his particular circumstances. It is not determined in any mechanical way by the commander's position. What is true is that the commander's position should imply that the local commander must make sure the roadblock is removed and give a general idea of the sanctions he can use: first, forcible removal, then adequate punishment of any retaliation on the part of the militia.

The linkage between these two levels is mutually reinforcing. If the commander took up his position and fallback position only at the level of his discussions with the militia leadership, without making sure his lower-level commanders implemented corresponding positions at their respective level, his position would be far less credible. Similarly the credibility of the local commander's position is enhanced by its being supported by the position of his commander.

Conflict Resolution Is Not a Matter of Taking Physical Action

Is not conflict resolution just a matter of the commander issuing orders to all local commanders to forcibly remove roadblocks and call in reinforcements if necessary to counter opposition?

The answer is no. Dealing with a confrontation is not just a matter of taking such-and-such actions, contingent or otherwise, upon the actions of others, although such actions may be a part of it. This is shown, in our case, by the fact that forcible removal of the roadblock is not part of the local commander's position in table 3. His position is that the local militia should remove it.

This is far from a trifling difference. Handling a confrontation is a matter of resolving a conflict through dialogue. The aim is that actions eventually taken should be willingly agreed to by all parties, ideally seen by each party as fulfilling its own objectives.

This would be ideal in the case we are examining. If the local UN commander gets his way in this manner, then the message going up from the grassroots to the militia leadership will be, "Agree to remove roadblocks! We're in favor." Forcible removal of roadblocks, on the other hand, would run the risk of increasing local opposition to their removal, or cause grassroots demands that the militia leadership resist this. It might thus work against the UN operational commander's mission objectives.

We are emphasizing, through this example, the difference between physical action, including the use of force, and resolving a conflict. The latter is not done directly through force, but through dialogue conveying the credible threat of force, leading, if successful, to genuine agreement.

The difference must be stressed because it is often obscured by the fact that dialogue itself may require physical acts, even large-scale ones, to demonstrate credibility. For example, warlike preparations, even up to beginning to use force, may be necessary to send the message that force will be used unless certain conditions are agreed to. But this is still dialogue, even though lives may be lost. It is dialogue because the aim is to send a message rather than directly to achieve objectives. This can be seen from the simple fact (Alberts and Hayes, 1995) that the element of surprise, always desirable when force is used directly to achieve objectives, must be absent. When forceful preparations, or force itself, are used to send a message, it is essential for the other side to know about them.

It appears then that force, although often an integral part of conflict resolution as carried out by armies, should be avoided if possible. To avoid using it, its use must be made credible.

What exactly is going on here?

THE SIX PHASES OF CONFLICT RESOLUTION

We need to explicate the place of force (or, more generally, actually realized conflict) in the process of conflict resolution. Figure 2 shows the general conflict resolution process divided into six phases. We will discuss these one by one.

Phase 1: Scene-setting

In this phase the problem to be resolved is set before the parties by the context, by a higher authority, or by a preceding confrontation. For a commander, this might be when he is tasked with a mission, or when new circumstances arise during a mission.

For the confrontation in table 3 we can be fairly specific. Scene-setting consists of the roadblock set up by the local militia and the receipt of information about it by the local commander. These are the events that bring them together in a confrontation.

The Informationally Closed Environment

As figure 2 indicates, scene-setting must logically set up a so-called "informationally closed environment" within which the confrontation may be resolved. This concept is important. Issues between parties can be resolved only on the basis of information available to them at the time, and this information must remain fairly stable during the course of negotiations. Resolution is impossible if new, relevant information continually upsets nascent understandings and commitments. Following are two contrasting examples

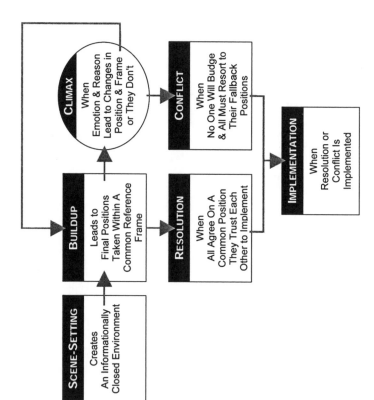

Figure 2. The six phases of conflict resolution.

that illustrate how demanding the requirement for informational closure can be:

- Recently a state came to a crisis because of persistent feuding between the private armies of two politicians. As a last resort, the army chief of staff took over and locked the politicians in a hotel room with phones cut off until they could reach agreement. In doing so, he enforced informational closure.

- During Operation Market Garden (the Allied drop on Arnhem in World War II), a reconnaissance pilot observed German heavy tanks in an area where British paratroopers were scheduled to drop. This indicated the presence of an SS panzer division; however, there was resistance to this genuine new information. Senior Allied staff had spent much time and effort agreeing to battle plans on the basis of the best available information. This naturally created a mind-set according to which conflicting new information should, if possible, be seen as incorrect. The alternative, after all, might require lengthy renegotiations of positions (in drama theory terms) and redrawing of plans throughout large organizations, which would have cost a great deal in terms of time, resources, and trust between players. The report was ignored.

The second example shows how informational closure, if taken too far, can have pathological effects. It nonetheless shows how necessary it is for conflict resolution, in this case, resolution of internal confrontations between Allied staffs. The point is that negotiations require stability, at some level, in the facts accepted as common ground between negotiating parties. The operational need is to resolve confrontations in a way that allows for flexible responses to new information, while preserving mutual confidence and understanding. Hopefully confrontation analysis will help us to meet this need.

Phase 2: Buildup

In this phase, dialogue takes place between the confronting parties to bring them into full confrontation. As figure 2 indicates, in this manner parties take up final positions within a common reference frame. This needs explaining.

The Concept of Common Reference Frame

Although communication between confronting parties does not require agreement between them, it does require a degree of mutual understanding. They must know the meaning of terms used by each other and know that each other knows that they know. The following analogies illustrate this: It is no use for me to threaten you with a gun if you think it is a popsicle. Nor can I bribe you with a popsicle if you take it for a gun.

The game-theoretic term *common knowledge* is useful here. It stands for what each party knows and knows that others know and knows that others know that others know…and so forth. Using this term, we can say that confrontation requires "common knowledge of the assumptions underlying communications between parties."

Without such common knowledge, they cannot adequately communicate threats and promises to each other; this requires that each must not only understand the other, but know that the other understands them, and so forth.

This essential set of communications assumptions shared between parties is their common reference frame. It is this set of assumptions that tables 2 and 3, showing confrontations with a militia force, are modeling. We consider a common reference frame to be specified by designating a card-table, together with an indication of how each player would rank the various possible futures in order of preference. These preference rankings can be conveyed by assigning players priorities for the various futures, as in our models.

The point here is that communication requires common knowledge of what the other is talking about (i.e., who the players are and what their cards are) and of each player's assumed preferences for outcomes (e.g., without common knowledge of preferences, what is meant as a promise may be taken as a threat, or vice versa).

Following are key points concerning common reference frames:

- **Simplicity is required for a common reference frame.** Only if a common reference frame is very simple can each party be sure that it is common knowledge (i.e., can really think, "I know that you know that I know...what I mean."). An example is when U.S. negotiator Richard Holbrooke confronted Bosnian Serb commander General

Mladic and Serbian President Slobodan
Milosevic in Dobranovci, near Belgrade, in
September 1995. Their common reference frame
was very simple. It contained two players, NATO
and Serbia. The cards under discussion were
NATO's continued bombing of the Bosnian Serb
army and the Serbian withdrawal of guns from
around Sarajevo. "This," said Holbrooke, "unless
that." All details that might distract from this
simple message receded into the background.
(See Silber and Little, 1996).

This could be considered an over-simplification
of a complex reality. The Bosnian government
and Bosnian Serbs were independent actors.
Milosevic recently had obtained a kind of
mandate to negotiate for the Serbs as a whole,
but on hearing Holbrooke's demands, he said,
"Come and tell Mladic yourself." Holbrooke
agreed; but Mladic was adamant. Holbrooke left,
threatening NATO action. Milosevic warned
Mladic, telling him that unless he agreed, NATO
would destroy his army. Mladic finally agreed.

A more complex model could be built; however,
any model can be made more complex by
including more detail. It also can be made
simpler by generalizing and omitting details.
Simple models are better representations of the
stripped-down common reference frames of
players at the climax of a confrontation than
ones that include distracting details. Such details
as those mentioned above would merely
illustrate our earlier point that a player's
decisions are determined by confrontations

between subplayers within that player (such as the face-off between Milosevic and Mladic).

• **A common reference frame may or may not represent parties' actual views.** This point too is important. If one party succeeds in deceiving another (e.g., by concealing the fact that it possesses certain weapons) then the common reference frame does not include the card, "Use those weapons," even though, if the truth were known, both parties would include that card in their considerations. Similarly, a party may deceive another by pretending to have a weapon it does not have, or pretending (falsely) to have the will to use a particular weapon. Going further, suppose that the second party is not deceived by some such pretense, but pretends to be. Then both parties will in fact share the same frame, while communicating in terms of a common reference frame believed in by neither of them. If these parties resolve their conflict, they do so within a common reference frame known by at least one of them to be inaccurate.

• **The common reference frame, being common, does not reflect the parties' differing values.** Each party generally considers itself to be in the right and thinks that others' values, and hence their way of describing the situation, are wrong. Thus they often use different terminology for elements of the common reference frame. What one calls "freedom fighters" another may call "terrorists" and a third "drug runners." A common reference frame, in our sense of the term, still exists, provided that all parties know what each other means, and hence know what each other is referring to using different terminology. Thus they are referring to the same players and cards (which for purposes of analysis we may name as we like, using neutral terms that best suit the party we are communicating with) and to the same preferences, even though different parties with different ideologies may give different reasons for and evaluations of these.

Final Positions

As figure 2 indicates, something more is required for the parties to build up (during the Buildup phase) to a full confrontation. The parties must take up positions and fallback positions within the common reference frame, as shown in tables 2 and 3, and these too must be common knowledge.

During the Buildup phase they may experiment with various positions to see how the others react. The Buildup phase comes to an end when the parties adopt positions they consider final within what they consider a final common reference frame. In the end, if the parties are to resolve matters, they must take a stand.

They then face what we call a "moment of truth," defined technically as a frame together with positions and fallback positions for each party.

At this point, one of two things must be true. Either all parties take the same position and, according to their common reference frame, can trust each other to implement it, or this is not the case.

Suppose it is the case. Then the parties have solved their problem as defined by their frame and their objectives within it. A more detailed model, or a model representing different issues, might show they have differences. Within this frame, they have none. As illustrated in figure 2, they therefore proceed next to the Resolution phase.

This does not necessarily mean the parties are being honest with each other. A solution according to their proclaimed positions and common reference frame may not be a true solution. One may be deceiving another. Each may be deceiving all the rest. If there is deception, it necessarily consists of each deceiver acting as if it has reached honest, trusting agreement. This is because successful deception requires the deceiving party to behave, not in accordance with the facts it believes to be true, but in accordance with the pseudofacts it wishes to be accepted. The behavior of such parties (to the extent that they are successful) therefore will be indistinguishable from the behavior of parties that are not deceiving each other, but genuinely believe in the version of reality they are projecting.

For example, the end period of the Nazi-Soviet Pact, initially signed in August 1939, may be a case of mutual deception. Although evidence of Soviet intentions is lacking (see Davies, 1996, pp. 1000-1013), it seems from their force dispositions that by June 1941 each was preparing a surprise attack on the other. The Nazis happened to strike first. Meanwhile, as far as their discussions went, the pact stood. This case actually takes the deception one twist further than we have described, in that not only did they deceive each other, they also disbelieved each other; however, each may have mistakenly believed that it was believed.

Phase 3: Climax

We have seen that one possibility at the end of the Buildup phase is that the parties find they have solved their problem, in that the dialogue between them records full, trusting agreement, even though, because of deception, the agreement may not turn out as expected.

What about the second case? Here parties either disagree on the terms of a resolution or openly distrust each other's intention to carry it out. They then move to the Climax phase as seen in figure 2. Here they face what is correctly called a moment of truth; something has to give. If it does not, they move into the Conflict phase, which consists of each undertaking to carry out its fallback position.

Fear of this outcome, which is in general not liked by one or more parties, puts pressure on them to change; therefore, something must give, and what must give is either their fixed positions within the given frame or the fixed frame itself. In other words, to avoid the Conflict phase, they must change one or more of the following:

- **Objectives**—Their own or others' objectives (encapsulated in their positive positions)

- **Threats**—Their own or others' implicit threats of what they will do if they do not get their objectives (encapsulated in their fallback positions)

- **Boundaries**—Their own and others' beliefs about the boundaries of their negotiations (modeled by the relevant set of players and cards)

- **Missions**—Their own or others' general aims pursued in the confrontation (resulting in the preferences they attach to various combinations of cards).

For example, look at the encounter described between U.S. negotiator Holbrooke and the Serb leaders. At the climactic moment of truth, the Serbs changed position. They agreed to withdraw their weapons. Did they also change their mission? Were they still determined to cleanse Sarajevo of Muslims? Most probably they were. The grudging nature of Mladic's change of position indicates that his concept of what the Serb mission should be did not change; however, Milosevic succeeded in overruling Mladic, and Milosevic may by then have had a different concept of the Serbs' mission. In that case, we would call the change a mission change as well as a change of position, although the mission change would be seen as precarious, depending on the balance of power within the Serb coalition.

Emotion and Reason

How then is change brought about? The answer is by means of emotion *(positive and negative)* and rational debate (or arguments in the common interest).

Holbrooke must have expressed the anger of the United States and the international community at the Serb bombardment of Sarajevo. (Note that ranting and arm waving are not the only ways of expressing anger.

An expressionless face in answer to another's emotion may suffice.) However expressed, such negative emotion would have helped to make it credible that he would bomb the Serb army into oblivion. Mladic, too, must have expressed anger, signaling his determination not to give in. At the same time, as a skilled negotiator, Holbrooke may have expressed positive emotion toward the Serbs contingent on changing their mission. The negative emotion of fear finally would have propelled the Serbian change.

Such emotions necessarily will have played an essential role; however, emotion on its own is not as effective as when it is supported by rational arguments in the common interest, based on logic, and if possible, including the production of evidence.

Holbrooke would have pointed out that neither side wished to see the Serbs destroyed. (Note that it was necessary to make this clear to the Serbians because, had it been part of NATO's mission to destroy them, the threat to do so would not have placed pressure on them to withdraw, because NATO's mission would have been to destroy them whatever they did.) For his part, Mladic may have tried to appeal to a supposed Serbian–United States common interest in opposing Muslim expansionism.

The function of rational arguments in the common interest, using logic and evidence, is to build credibility. Mere emotion can appear to be not serious because its effect may be considered transient; however, if I can convince you that my position and fallback position, in addition to being invested with strong positive and negative emotion, also are based on logic and evidence and are in the common interest, they become very credible. This is because offering reasons makes it harder for me to give in and easier for you. Often, neither reason nor emotion, on their own, are credible motivators; together, they are.

Rational common-interest arguments also tend, if successful, to weld the confronting parties together into what is seen as one party, pursuing a unified set of interests. This is the converse of the fact, already noted, that the preferences and attitudes of a player are determined by confrontations between the subplayers composing that player.

Subplayers, by continually solving their subconfrontations within a player, in effect continually construct the player to whom they belong as a purposeful entity able to take decisions and, through its representatives, express emotions. Similarly, when any confrontation is properly resolved, good feelings between the parties and sentiments of loyalty toward the agreement reached are created. If organizational requirements also are met, these feelings tend to create a single superplayer comprised of the original players.

Some further points need to be made about emotion and reason:

- **Warning and Notification**—If I do not need to change my preferences to make my threat or promise credible (because they are things I would want to do anyway), my threat is more like a warning and my promise more like a notification. To convey this, I may show lack of emotion.

- **Deception**—Conversely, to deceive you into thinking that my threat is merely a warning, my promise a mere notification, I may show lack of emotion. For example, Holbrooke may have tried to show the seriousness of the U.S. commitment by deliberately showing no emotion.

• **Irrationality**—If I have no good reasons for preferring to reward or punish you, I may make my promise or threat credible by strong positive or negative emotion accompanied by evidence of irrationality so that you think, with delight or horror, "He is mad enough to do it!" This is more likely to be effective, the more immediate is the possibility of implementing the threat or promise because we know that emotions often are temporary. For example, your anger is more likely to be effective if your guns are pointing at someone than if it will take 3 weeks to transport them from the United States.

Phase 4: Conflict

If a change in positions or common reference frame takes place in the Climax phase, it requires a return to the Buildup phase, as shown in figure 2. This is necessary to communicate the new configuration of positions within the new reference frame and to establish that they are common knowledge.

It is possible change still may not take place. Emotions and the search for reason and evidence may not be enough to overcome inescapable evidence (e.g., Hitler has invaded Poland), unforesakeable values (e.g., Europe should be democratic rather than Nazi-ruled), and unshiftable positions (e.g., Poland shall not be occupied by Germany; if it is, Britain will declare war). If change is impossible, by assumption the characters must resort to their fallback positions; that is, move to the Conflict phase (e.g., World War II). Here the parties must prepare for what we have called the threatened future, the appropriate term for the conflict as imagined or foreseen by the parties.

At this point, the parties split up (their bid to achieve their position having failed), and each decides if it now intends to implement the fallback position it has implicitly or explicitly committed to. This phase is a serious decision point. Internal subplayers belonging to each player now may confront each other, some demanding to carry out their commitment, others unwilling to do so. Estimates on what other parties will do are important to a party making this decision. Will they carry out their commitments to the threatened future? What alternative courses of action will they take?

These internal confrontations also can be modeled by the card-table method, using models linked to the original overall table.

In Athens in May 1993, United States and British envoys Cyrus Vance and David Owen, backed by Bosnian President Slobodan Milosevic, persuaded Bosnian Serb leaders to sign the Vance–Owen plan for dividing Bosnia while keeping it as one nation. According to Vance's right-hand man: "Vance suddenly tells them straight out that this is the last-chance café, the U.S. Air Force is all prepared to turn Bosnia and Serbia into a wasteland." This was the threat Milosevic had been hammering home to them. So the Bosnian Serbs signed, subject to the approval of its parliament, which rejected the plan. For a time after, U.S. officials tried to persuade Europeans to agree to implementation of their threat. They failed.

This United States–European subconfrontation ended with the United States shifting position and agreeing not to anger its NATO allies by "act[ing] alone in taking actions in the former Yugoslavia." There was a concomitant change in United States values (i.e., the values it was pursuing in former Yugoslavia), initiated by President Bill Clinton and his wife Hillary, who read a book that persuaded them the Balkans were doomed to violence. We interpret this as follows: In the Climax phase of the West-versus-Serbs confrontation, the Serbs decided to call the West's bluff. The players then moved to the Conflict phase, in which the West's decision (through the resolution of confrontation between the United States and Europe) was to not do as they had threatened to do (Silber and Little, 1996, chapter 21).

Thus the Conflict phase may end in an Implementation phase in which the threatened future is carried out or one in which it is flunked. There is a third alternative. There may be a return to the Climax phase if one or more parties decides, as a result of internal confrontations, that after all they are prepared to change, and therefore request to reopen negotiations. They may then cycle back from the Climax phase to a renewed Buildup phase.

This happened in the example of the Holbrooke–Milosevic–Mladic confrontation of September 1995. When Holbrooke left the meeting, negotiations moved into the Conflict phase. Milosevic and Mladic then had a subconfrontation as subplayers within the player Serbs. This was resolved by Mladic caving in. The Serbs then called Holbrooke back.

The Use of Force

In this example, the Western threat was to use force. Is a threatened future necessarily one of armed conflict?

At the political level, a threatened conflict may not involve force at all. Economic sanctions, for example, may be the only threat. The mere involvement of the military tends to carry the implicit threat of the use of arms, even though not necessarily their direct or immediate use.

In classic United Nations peacekeeping operations, if former combatants reject their position in favor of a continuing cease-fire, the fallback position of the United Nations is simply to withdraw UN troops. There is no theater-level threat of United Nations use of force, although this is threatened at lower levels if there are attacks on UN troops, who are supposed to defend themselves if attacked. Although the UN does not itself threaten to use force at theater level, UN withdrawal may lead to resumption of armed conflict between the parties, as happened in May 1967, when the withdrawal of UN troops was followed in June by the Six-Day War (Alberts and Hayes, 1995, p. 18).

Phase 5: Resolution

If parties are to finally resolve their differences, rather than fall into Conflict, according to figure 2, they must cycle between phases of renewed Climaxes and renewed Buildups until they converge on a common position within a frame that ensures them of each other's trustworthiness. Each time they cycle through the Climax phase, they risk moving into the Conflict phase. If they successfully resolve their differences at the Climax level, they move to the Resolution phase.

After the parties sense they have reached an understanding sufficient to resolve their problem, they must communicate to give mutual reassurance that each understands what is required. This takes place in the Resolution phase. Here details are discussed and understandings translated into more or less unambiguous commitments.

It is always possible that this consideration of details and firming up of an overall understanding may reveal flaws sufficient to differentiate the parties' overall positions or cause serious mistrust. That will necessitate a return to the Buildup phase followed by a renewed Climax phase, with the renewed danger of conflict. To avoid this, the detailed discussions need to be carried out in an atmosphere of mutual goodwill (positive emotion) to encourage parties to see each others' viewpoint, find win-win solutions, and trust each other; however, this general atmosphere cannot mean that detailed differences are not explored. That is the reason for this phase. Exploring differences gives rise to negative as well as positive feelings. The challenge in this phase is to avoid going back to the Buildup and the Climax phases by keeping the overall emotional tone positive while allowing negative feelings to grow up and die down, but never take over.

Even after Germany's unconditional surrender in World War II, it was necessary for the Allies to sit down and negotiate detailed issues of governance with German officials. After receiving Japan's surrender, British officials in some former colonies had to negotiate the continued temporary employment of Japanese troops as a police force. The overall solution, unconditional surrender, did not determine all these details. Nor did it determine what the implicit fallback positions of German or Japanese officials were in these negotiations. It must have been non-cooperation in various degrees. What if this non-cooperation had been multiplied and increased by general failure to negotiate satisfactory solutions? This is what must have happened in the negotiations between German invading forces and the at-first-welcoming inhabitants of the Ukraine and other areas of the Soviet Union. We suppose the result was first a return to the Buildup phase where the position taken was "Treat us better." In the immediately ensuing Climax phase, the Ukrainians came to a realization that this position would be rejected. Finally, the Ukrainians reached a Conflict phase where they decided to continue the war with armed resistance.

Note that discussions at the Resolution phase need not solve all the problems raised. By recursion, that would mean all details had to be solved before outline agreement could be reached, which is absurd (e.g., agreeing the U.S. Constitution did not require settling all political differences between parties). For many problems, discussions will result in mere agreement on parameters for future interactions. The Resolution phase looks at details solely to confirm the agreement reached within the simplistic, overall, common reference frame arrived at following the last Climax phase.

Modeling Considerations

Each difference or group of differences emerging in the detailed negotiations of the Resolution phase can be modeled by a submodel, where differences are represented by differing positions. These submodels then can be linked.

At one extreme, each separate issue would be modeled separately. At the other extreme, all differences would be brought together in a large card-table. (Note that there are no practical limits as to how many cards you can put in a card-table model.) The chosen modeling procedure should reflect a decision (which itself can be guided by building different models) as to whether to divide issues into separate groups to solve separately, or to solve them together. Solving issues together (i.e., modeling them in one card-table) allows consideration of trade-offs between issues. Modeling them in separate but linked card-tables does not allow consideration of trade-offs (i.e., it is difficult to model an agreement that consists of a quid pro quo between the issues represented in one card-table and those represented in another). Separate but linked consideration does allow the effect of issues on one another, inasmuch as reaching a certain agreement in one card-table may have an effect, through causal linkages, on players' preferences and the cards they are able to play in another, linked card-table.

Phase 6: Implementation

We have seen several possible kinds of Implementation of what is decided in a conflict resolution process. What is implemented may include any of the following:

- **Resolution of the confrontation**—What is implemented is the common position of all parties in their common reference frame, with details having been confirmed in the Resolution phase (e.g., the overall agreement between parties in Northern Ireland reached during Easter 1998)

- **False resolution**—A common, agreed position exists but is not implemented because parties try to deceive others (e.g., the end of the Nazi–Soviet pact in 1941)

- **Conflict**—Implementation of the threatened future, where all parties decide in the Conflict phase to implement their fallback positions (e.g., Britain's 1939 declaration of war on Germany)

- **Flunked conflict**—Resolution has not been achieved, and the parties declare the threatened future should be enacted; however, it is not enacted because one or more of the parties decides not to implement its fallback position (e.g., the 1993 aftermath of Serbian rejection of the Vance–Owen agreement).

These are four ways a confrontation may end. A fifth way is by interruption (i.e., by a breach of the informational closure that stabilizes the assumptions underlying the confrontation). If the parties receive new, relevant information from exogenous sources that changes their assumptions and expectations as created in the Scene-setting phase, then all bets are off. If the parties to the present confrontation continue to interact, it will be through conducting or resolving a new confrontation, one that starts with a new scene being set (e.g., a new confrontation started when the

Japanese attack on Pearl Harbor changed the assumptions underlying the confrontation between the United States and British parties as to whether the United States should enter the war. The confrontation was speedily resolved in favor of United States entry into the war).

Implementation and the Unexpected

Interruption, by definition, creates an unexpected future. The same generally is true of the other four ways a confrontation may end. These other four endings attempt to implement a known future, but this attempt generally fails because it contains surprises. We cannot foresee the future.

This is more than a matter of uncertainty (a known set of possibilities with known probabilities attached to them) or risk (a known set of possibilities with unknown probabilities). Uncertainty and risk apply to the future as projected by parties beforehand. The future that is actually realized is usually one the parties did not think of as a possibility.

The lack of foreseeable future is well known to warriors. As Clausewitz emphasizes, parties' projections of what will happen if a threatened future is one of armed violence usually fall wide of the mark. The point for confrontation analysis is that these projections, not the reality, are what persuade players to choose one path or another.

Unexpected Contingencies and the Need for Positive Feelings

Agreements between parties are necessarily contingent on the assumptions about the future the parties made at the time, even if they know these assumptions will turn out wrong. They know their agreement will have to be reinterpreted by each of them under circumstances not foreseen in the agreement. This means that each agreement involving future cooperation between parties needs to contain a general card for each one to interpret this agreement with the other's interests in mind. To make the playing of this card credible, a general cementing of feelings of long-term goodwill or love is needed between parties. Such feelings are well known to warriors. A platoon going into action needs each member to credibly play the card, "I'll risk my life to save yours, whatever happens"; hence the love between soldiers. Similarly, cooperation between different national contingents in a combined command requires the buildup of positive feelings.

REVERSIBLE AND IRREVERSIBLE DECISIONS

Every event is irreversible, in one sense. If it happened, it happened. Yet the logic of the conflict resolution process as shown in figure 2 denies this. To resolve or try to resolve our differences requires a time and space where the only things that happen are like the putting down and picking up of cards. They are seen as reversible. We need to be able to say, "They said that, but we can change their mind. We can reverse what was said." This time and space, created in the

Scene-setting phase by informational closure, is a place where everything is reversible because everything is seen primarily as a message, symbolic of what we intend to do, rather than something done for its own sake. The logic follows this thought: *Nothing is irreversible until the end* (i.e., until one of the five endings).

This means that until the Implementation phase, or until the conflict resolution process is interrupted, all that can happen is communication of intentions, beliefs, values, and reasons.

Holbrooke walked out on Milosevic and Mladic at their meeting in September 1995, thereby taking the confrontation to the Conflict phase, but he was able to resume talks with them when they relented, thus moving back again to the Climax and from there to a new Buildup.

Because nothing irreversible can happen during the first five phases, any cards whose playing is by nature irreversible cannot be played until the Implementation phase. This rule is merely a matter of definition. It means that a significantly irreversible decision would be taken to represent the beginning of a new confrontation, different from the old one inasmuch as that decision would now be irreversibly fixed. By the same token, if a card is by nature reversible, then an important, even necessary way of signaling an intention to implement it may be to start doing so.

The NATO air strikes that accompanied the same Holbrooke–Milosevic–Mladic meeting had, of course, irreversible effects in damaging Serbian equipment and taking Serbian lives; but their real function was to signal NATO's conditional intention to continue attacking the Serbs. This is shown by the need to let the Serb leadership know the strikes were going on, and why, even though by the principle of surprise (Alberts and Hayes, 1995, pp. 29-30), the attacks would have been more effective in destroying the Serb's war-fighting capacity if they had been unaware of the strikes or the intention behind them.

The air strikes' function as a signal depended on their being reversible in the large (i.e., on the possibility of stopping them if the Serbs relented). Of course, if continued long enough, the air strikes might have caused so much damage as to create an irreversibly changed situation, a fact that set a kind of deadline for the confrontation to be resolved. Yet the air strikes were, by nature, sufficiently reversible that the signal, "We'll bomb you," had to be, at a certain point, reinforced by starting to do it or it would not have been credible. The point at which this need for credibility kicked in was partly set by the West's previous undermining of its own credibility.

By contrast, the Cold War nuclear threat, "We'll take out Moscow," could not be and was not expected to demonstrate its credibility by being carried out. It was too irreversible.

You Can Destroy a Parking Lot Only Once

There is only one Moscow; hence by carrying out a threat to nuke Moscow, you lose a card from your hand. You have one less means of exercising pressure.

As an example, there is a possibly apocryphal story that tells of a Bosnian Serb leader who was opposed to a cease-fire receiving a phone call from NATO. "Look out the window. See your parking lot? With your Range Rover in it? That noise is a British jet. It'll be with you in two minutes. Know what's in its sights? Your parking lot. Now perhaps you'd like to change your mind... The General? He is standing beside me." The Bosnian Serb changed his mind. The point of the story is that if the parking lot had actually been destroyed, such pressure could no longer be placed on the Bosnian Serb leader. Destroy his house, his family, and everything he values, and he has nothing left to lose. Consequently it may be impossible to put any pressure on him to concede.

This contrast between destroying other-party assets to win a battle and to win a confrontation could not be starker. Each asset I destroy increases my relative physical strength but loses me a bargaining card; therefore, it is counter-productive to destroy other party assets when the following situations exist:

- **I have sufficient preponderance of physical force not to need the physical strengthening I get from destroying his assets.** In general, the United States and its allies have this kind of preponderance in the post-Cold War world, where we are typically "the strong fighting the weak"; nevertheless, guaranteeing the physical security of our forces may require selective destruction of assets.

- **I don't need to start destroying his assets to make it credible that I'm prepared to.** This refers to the message-sending function of asset destruction. The problem I may face is the need to use up some of my bargaining assets to make the rest credible. This problem may be solved by minimizing actual destruction while maximizing its psychological effect and, at the same time, sending the right messages to accompany it.

WAR-FIGHTING THAT IS NOT MESSAGE-SENDING

We should stress that the discussion in this chapter has concerned a military campaign whose purpose is to conduct a confrontation and, if possible, resolve it on our terms. In this context, the primary purpose of much of the military action that takes place is to send a message; however, this describes only one kind of campaign, the kind usually called an OOTW, although the threat of war normally forms part of it, at least in the case of peace operations. Peace is kept by credibly threatening war. This is why such missions are undertaken by the military.

In chapter 1, we argued that such missions are increasingly common in the post-Cold War world; however, military missions of the traditional, Clausewitzian kind also occur. Here the mission is simply and directly to destroy the military assets of the other side, thereby lessening or eliminating their ability to do us harm. Message-sending has little to do with it.

The purpose of Clausewitzian war-fighting is, in general, to set the scene for a subsequent peace in which the victor can dictate terms, having achieved a monopoly of military force. As Clausewitz says, while stressing that war is always a means to a political end: "Whatever may take place subsequently, we must always look upon the object as attained, and the business of war as ended, by a peace." Clausewitzian war is undertaken, not primarily to send a message, but to change the facts on the ground so that subsequent conflict-resolution processes may have a more favorable outcome.

We can fit direct, purely military action into our model. It can be seen as belonging to the Scene-setting phase of figure 2. War of this kind is launched by politicians who foresee that if they and others enter negotiations equipped with the cards they presently hold, the result will be unfavorable to them; therefore, they act to take away certain cards held by others, cards consisting of effective military response. Surprise and other generally accepted principles of war apply fully.

Preemptive action of a traditional military kind also may be taken to interrupt a process of conflict resolution to be able to start another on more favorable terms, after others have been deprived of certain cards. This may happen in two following ways:

- The aggressor may intend from the beginning to take the preemptive action. It may simply use the process of confrontation and conflict resolution to lull others into lowering their defenses.

- Alternatively, the aggressor's participation in conflict resolution may be genuine up to a certain point. It then receives new, exogenous information, the result of which is to raise its demands above what it can reasonably expect others to accept, given the present distribution of cards among the players. So it decides to forcibly take certain cards from others' hands.

Hitler's decision to interrupt discussions with the Soviet Union and invade Russia may be an example of the latter kind, looked at from a short-term viewpoint. In a longer-term perspective, it was probably a case of the first kind. He always intended to attack the Soviets eventually.

Another example might occur if Iraq were discovered to possess a nuclear bomb, and the Allies knew its location. To prevent him threatening the West with it, the decision might be made to destroy the weapon by a commando or air raid. If successful, this would remove the card, "Use nuclear weapon," from Saddam's hand.

These examples are Clausewitzian in that war is being waged for its physical effect in removing another's ability to harm us. Clausewitzian war may, of course, also be waged by a defender of the peace in reaction to an act of aggression.

We are not seeking to say anything original about traditional warfighting, merely to contrast it with OOTW and to point out the essential difference. Whereas traditional war-fighting merely precedes and sets the scene for the message-exchanging activity of conflict resolution, an OOTW consists in large part of this message-exchanging activity itself. It consists of conducting a sequence of confrontations and attempting to resolve them in such a way as to resolve the issues in accordance with our objectives. That is why the primary function of most actions taken in OOTW is to send a message.

SUMMARY OF CHAPTER 2

An example of a UN force dealing with roadblocks showed how confrontations at different command levels are linked together. Handling them in a coordinated way reinforces the effectiveness of messages sent, and we point out that handling confrontations is essentially a matter of message-sending, not physical operations.

Conflict resolution is a natural human process divisible into six stages: Scene-setting (when the problem is posed within a given context); Buildup (when players take positions and fallback positions within a common reference frame); Climax (when they must either redefine their positions and their frame, or else fall into conflict); Resolution (by having changed they find they agree and can trust each other to implement their agreement); Conflict (when having failed to agree, they are faced with having to carry out their fallback positions); and Implementation (when they carry out or decide not to carry out either the agreement they have reached or the conflict they have fallen into).

A common reference frame embodies the minimal set of assumptions players need to share to communicate. It needs to be simple, it may be deceptive (because players can deceive each other), and it is interpreted differently by players with differing values.

At the Climax, players use emotion and reason (positive and negative) to change themselves and one another. Reason and emotion are each relatively ineffective without the other. Arguments based on the players' common interests are most effective; consequently, players conducting confrontations tend to build up the preferences of a superplayer to which they all belong.

Conflict may be flunked (i.e., not implemented because of unwillingness to carry out threats that have been made), just as agreements may be betrayed. Because the future cannot be foreseen, conflict outcomes in any case rarely turn out as projected, and agreements need goodwill between participants because they will need to be maintained in circumstances not foreseen when they were launched.

During conflict resolution, nothing irreversible can be done, because all that is supposed to happen (until the Implementation phase, provided the process is not interrupted) is dialogue. Credibility may require beginning to carry out threats, but that is still regarded as a way of sending a message. There is a problem when we send a threatening message by destroying assets. This defeats the purpose of the message because the threat does not exist when the assets have been destroyed; therefore, it may be best to destroy low-value or replaceable assets provided this has sufficient psychological impact.

CHAPTER 3

HOW PARTIES PRESSURE EACH OTHER: COOPERATION AND TRUST

Our continuing aim is to show how a commander can form a strategy to win a confrontation in the sense of getting his position genuinely accepted.

Of course, the most excellent strategy conceivable cannot by itself guarantee victory, any more than it can in battle-fighting. Superior strength and morale are also important, just as they are in battle-fighting, although in a different way. There is a further complication in that mission objectives themselves may change in the course of a confrontation, and with them the very definition of success; nevertheless, it is important, as in warfighting, to have the best possible strategy for obtaining given objectives. How is this developed?

In the confrontation shown in table 2, how should the commander work out a strategy for getting the militia to comply with the agreement to cease ethnic cleansing and allow free movement? How should he devolve his strategy to other levels of his command, such as the level of the commander in table 3?

To answer these questions, one part of the resolution process is of prime importance. It is the Climax phase, as shown in figure 2, where changes take place in characters' attitudes. In this chapter we analyze these changes and show how they are brought about.

Understanding them is the key to success. It must be, because the commander requires these changes to go in the direction of his objectives.

As we discuss below, the commander's job as a peace-operations professional is to build on the fact that there is a tendency for changes to occur unless and until a full resolution is reached (i.e., until all parties have converged on a single solution they can trust each other to implement). He must direct and orchestrate these tendencies to ensure that resolution is achieved as close as possible to his position.

GAMES VS. DRAMA

Therefore we take up the question, "What drives the changes in attitudes, beliefs, and objectives that take place in the Climax phase shown in figure 2?

To lay the foundations for an answer, we will look at the main assumptions underlying von Neumann's theory of games and the more general approach within the social sciences that calls itself "rational choice theory" and reaches its fullest development in game theory. (For game theory, see von Neumann and Morgenstern, 1959. For a clear account of its present reformulation, following decades of development, see Osborne and Rubenstein, 1994.)

Our actual question is debarred by game theory. The subject defines itself in such a way that the question, "What drives these changes?" cannot be asked. It cannot because rational choice theory has a mission. It tries to take the idea as far as it can that all choice is rational; and the term *rational* is used to mean "pursuing fixed, given preferences within a fixed, given frame." To assume rationality in this sense is to assume that the kind of change depicted in figure 2 is impossible. Characters' beliefs about the frame and their preferences for possibilities within it (although not their positions) are assumed to be fixed. Of course, exogenous changes in these beliefs and preferences (interruptions) brought about by new information coming from outside, are allowed, just as we allow them. But change brought about merely by characters pressuring each other is assumed not to occur.

Why do we discuss this rationalistic view taken by game theory if it excludes the possibility of answering our question? One reason is that the reader may have come across it. It is important throughout the social sciences, it dominates our understanding of economics, and it is influential in political science, negotiation theory, and military theory.

That is not our main reason for discussing it. The main reason is that in pursuing its limited definition of rationality ("the pursuit of fixed preferences within a fixed frame") game theory uncovers numerous dilemmas, the best-known being "prisoner's dilemma" (see in particular the early discussions by Schelling, 1960; Rapoport, 1964, 1966; Howard, 1971). These dilemmas show in various ways that the pursuit of fixed preferences within a fixed frame is not what it seems; to achieve outcomes high on one's list of preferences, it may actually be advantageous to forego pursuing one's preferences because characters generally need to make threats and promises they would prefer not to carry out.

- The West wanted the Serbs to stop invading Bosnia. To achieve this, the West had to threaten to do something it preferred not to do, attack the Serbs. Fixed, known pursuit of the latter preference (not to attack) prevented satisfaction of the former (Serb cessation of violence).

- The Serbs did not want to be bombed. They also wanted to take over much of Bosnia. When the West made its ultimatum credible, Serbian pursuit of its preference for taking over Bosnia prevented it from satisfying its first, greater preference, cessation of bombing.

Game-theoretic dilemmas are created by a tension or contradiction between the values a character wants to pursue and the credibility it (the character) needs to have to pursue those preferences. The dilemmas so created are dilemmas not just for the theorist, but for the players themselves. They are in paradoxical situations: the prisoners in prisoner's dilemma actually see that if each chooses so as to guarantee the worst possible outcome for itself (given the other's choice), both do better for themselves than if each guarantees the best possible outcome for itself (again, given the other's choice). (Howard, 1971, p. 45).

This illustrates that game-theoretic dilemmas bring players face to face with psychologically intolerable paradoxes and gives us a clue to explaining the changes that take place in the Climax phase of a confrontation. We make the following hypothesis: players make changes in their common reference frame, their own and others' preferences, and the purportedly final positions they have taken up in an attempt to eliminate paradoxical dilemmas. We hypothesize, therefore, that emotions are aroused in a predictable way by game-theoretic dilemmas. A player is pointed in the direction of dilemma-eliminating changes. Reason and evidence are invoked to justify, if possible, changes that emotion has pointed to.

For example, Western public opinion eventually became sufficiently emotionally aroused against the Serbs to eliminate the dilemma the West faced, that we preferred not to attack the Serbs even though we needed to make credible our threat to do so. Our anti-Serb feelings, aroused by this dilemma, were rationalized by demonizing the Serbs (i.e., by propagating the idea that they were evil and supporting it by selectively highlighting the atrocities they committed while ignoring those committed against them). After we achieved such rationalization, our preferences changed. We now preferred to attack the Serbs rather than let them continue attacking Bosnia.

The proposition that emotion evokes the use of reason and evidence to try to bring about dilemma-eliminating change is the basis of the discipline, drama theory that underpins confrontation analysis. (See Howard, Bennett, Bryant, and Bradley, 1992; Howard, 1994, 1994a, 1996, 1998; Bennett and Howard, 1996; Bryant, 1997. See also Nigel Howard Systems, 1992–1997).

It is, perhaps, a disconcerting proposition if one believes that reason and evidence should be used to try to reach objective, unbiased conclusions rather than to prove a point given beforehand. This belief is not generally well founded. We find the very epitome of objective reasoning in scientific method, the closest method for arriving at objective truth. Even so, students of scientific method stress that scientists too must have a hypothesis to prove or disprove before selecting evidence or using reason (see Popper, 1959; Kuhn, 1962). The important thing about the collocation of reason and evidence is not that it is impartial or disinterested—in the sense of having nothing to prove. That is not the case. Its importance is that, if effectively presented, it compels belief in what it aims to prove.

That is precisely its function in a drama-theoretic confrontation. Emotion may make me heard; I need reason and evidence to make me believed. The reason I need to be believed is that I need to make my position credible.

Why the Drama Metaphor?

The word "drama" replaces the word "game" in the theory that underlies our approach. Practical users need not bother with fundamental theory; nevertheless, they may ask why. What does the term "drama" point to that "game" obscures? Both games and drama are role-playing, leisure activities that are used to illuminate life itself. The game metaphor was used by the military for training purposes long before game theory was conceived. Do we need a change of metaphor to understand the basic nature of OOTW?

To see that we do, compare behavior in game-playing with behavior in drama. A game lays out a fixed set of choices (sequential or simultaneous), specifies what outcomes (stochastic or deterministic) to expect from each combination of choices, and requires players to have specific preferences over outcomes. It is forbidden either to vary your preferences (e.g., prefer another to win, or to invent other choices, such as move another's pieces). The result is that as long as actors play the game, they have nothing to do but predict what others will choose and choose the most-preferred outcome for themselves, given their predictions. This is precisely what game theorists mean by rationality.

By contrast, in a drama we see characters being emotional and irrational and participating in rational debate. In doing so, they pass through a crucible that changes them. Both their value systems and their views of reality change. The very rules of the game change.

The chief interest of drama is in seeing how this happens. The main thing we learn from drama is not, as with games, how to be rational in the sense of pursuing fixed preferences against others with fixed preferences, all the time keeping to fixed assumptions about what is possible. Instead, we learn how players (now called "characters") in interaction with each other use reason (in another sense of the term) and emotion to change their common assumptions and to work on the value roots of their own and others' preferences to change them also.

The way a drama ends differs from the ending of a game. A game ends with the victory of one side. Losers are left in an artificial state of frustrated discontent. It is artificial in the sense that only the arbitrary rules of the game prevent the losers from doing anything about it. If they choose to operate outside these rules, they may resort to throwing the pieces on the floor, wrecking the grounds, or attacking their opponents.

By contrast, a drama ends when none of the characters has anything left to hope for or to fear. This characterizes a state of stability, with no further tendency for characters' expectations to change, although new external information (interruptions) may disrupt this state. In a real-world drama, the dénouement never turns out as expected, a fact passed over in fiction by not showing what happens afterwards as a result of the agreements and understandings reached.

The end of a drama is nevertheless characterized by complete stability of expectations, which may be brought about in two different ways represented by tragic and happy endings. A tragic ending is stable because characters' fears have been realized and their hopes destroyed; a happy ending is stable because characters' hopes are realized and fears banished. In either case, they no longer have hopes or fears.

From a military commander's viewpoint, what matters is that the ending of a drama represents stability, while the ending of a game does not. Losers of a game look forward to another round, if they do not demand a replay or resort to violence. Characters at the end of a drama are content, seeing no alternative to acceptance of what they have.

Such contentment, enforced or willing, must be the commander's object in an OOTW. The parties need to be brought into a state of contentment with a future that consists of fulfillment of the commander's mission. They may, of course, be in various states of discontent and conflict in regard to other matters, particularly details. The aim of a peace-support operation is generally to bring about a broad degree of peace and order, not to settle every issue between inhabitants of a region; nevertheless, in regard to the broad societal issues that he has to settle, the commander's object must be to make the parties content with his position.

The means for obtaining that kind of stability through contentment are examined in drama theory.

The Need for Game Theory

We must stress that despite its willingness to violate game theory's basic assumptions, drama theory needs game theory. In some places it needs game theory directly. When for any reason characters cannot interact with each other to negotiate a resolution (i.e., when preplay communication is disallowed) drama theory reduces to game theory.

This, in particular, is the state of affairs when parties enter the Conflict phase as shown in figure 2. At this point they have failed to reach agreement and must decide whether to implement the fallback positions they are committed to implementing or, if not, what else to do. They must make this decision having cut off communications with each other—unless, that is, they decide to go back to the Climax phase by asking to reopen negotiations on the ground that they have a change to announce. Failing this, the calculations they must make include hypothesizing if the others will stick to their fallback positions. If so, what should I do? If not, what else will they do? What will they think I am going to do? All these responses are game-theoretic.

This is true only of interactions between the characters as such. Internal confrontations between the subcharacters making up a character generally continue to be drama-theoretic. Even so, it is clearly one way in which game theory is part of drama theory.

Warfighting itself is essentially game-theoretic. In the traditional Clausewitzian model, the military takes over when politicians fail to resolve a confrontation in which Conflict (i.e., the Conflict phase as shown in figure 2) means war. In such case, the development of a military strategy to destroy the enemy's war-fighting capacity is essentially an application of game theory, regardless of whether quantitative game theory (the form in which it is usually presented) is considered useful. On the other hand, internal relations between the different units of a joint or combined force are drama-theoretic.

Dependence on Game-Theoretic Dilemmas

A fundamental reason drama theory needs game theory is that it uses the dilemmas produced by game theory to analyze, predict, and understand the pressures on characters in a Climax phase to change their positions, preferences, and common reference frame. The following logical derivation of drama theory shows how.

- Initially we suppose that characters see their common reference frame as fixed, and hence see themselves in a game. This is because they have assumed this frame for purposes of communication and for the time being, cease to question it.

- As the characters see themselves in a game, they try to behave rationally in the game-theoretic sense. This brings them up against game-theoretic dilemmas.

- The dilemmas then cause the characters to feel positive and negative emotions. The emotions felt depend on the dilemmas the characters face and the ways they deal with them.

- Emotions may cause them to behave irrationally (i.e., not optimally), change their preferences, or search for new cards to play or new characters to introduce. By these means the dilemmas may be eliminated.

- As an alternative, the characters may solve their dilemmas by changing their positions.

- A general alternative to these methods of dilemma-elimination is deception. Instead of becoming irrational or changing preferences, cards, or positions, characters may pretend to do so. Note that deceitful persuasion changes the common reference frame just as much as non-deceitful persuasion.

- What makes deceit attractive for one character creates disbelief in another. Such disbelief must be overcome to effect the desired change in the common reference frame. To overcome it, characters construct logical arguments, show evidence, and appeal to generally accepted standards.

- Characters' arguments cannot be value-free; to make sense, arguments generally must assume the pursuit of some common interests or objectives. If objectives are not shared, arguments based on them will seem insincere or unappealing. Characters are thus led to construct rational arguments in the common interest.

- Successful arguments of this kind have the effect of building up the preferences and attitudes of a supercharacter formed by an alliance of the characters. Individual characters' preferences do not become the same as the supercharacter's, but become such that solving subcharacter confrontations becomes a mechanism causing the supercharacter to function as a character.

- The supercharacter generally will be a character in a larger drama, in which it too tries to behave rationally and so confronts dilemmas.

Starting from the assumption that characters try to be rational within a fixed frame, drama theory shows how they are led to behave irrationally and change the frame, so creating the possibility of rational behavior at a higher level. This is the deep sense in which drama theory depends on game theory.

THE COOPERATION DILEMMA

Having seen in general how the encounter with game-theoretic dilemmas leads players to change, we look in detail at how this happens.

There are six dilemmas that can place pressure on characters: cooperation, trust, deterrence, inducement, threat, and positioning. We will examine each and discuss the kinds of change necessary to overcome it. Each dilemma is illustrated with examples, including a reference to the roadblocks-removal problem in tables 2 and 3.

The cooperation dilemma faces a character that is tempted to defect from its own position. (This arises because my position generally will contain things I do not want, but have included as concessions to others. I may be tempted not to carry them out).

A cooperation dilemma faces a character at a card table when, by changing its own selection of cards, the character can move from its own position to a future that it likes just as well or better. Table 2, column P, illustrates. Here, the ethnic militia has accepted the commander's position, thus also adopting that position for itself. Actually, the militia prefers the future (not shown on the table), otherwise the same as P, in which it does not play the card, "Allow free movement." The militia can move to this future (if the commander carries out his part of P) merely by changing its own selection of cards; therefore, the militia's own preference gives it a cooperation dilemma.

The cooperation dilemma can be thought of as the "drunk's dilemma." Imagine an alcoholic whose wife threatens to leave him if he will not stop drinking. He does not want her to leave, so he swears he will stop. He knows, and knows that she knows, that if she stays he will not be able to keep his promise, but will contine to drink.

General Statement of the Cooperation Dilemma

Generally stated, the cooperation dilemma faces a character that belongs to a group (subset) of characters, all of whom can, by changing just their own card selections, move from the character's position to another future they all like just as well or better.

For example, assume the drunk depends on a visiting friend for his supplies of booze, as shown in table 4.

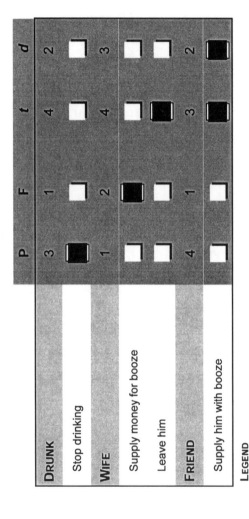

LEGEND

■ means card is played

□ means card is not played

P is position of drunk and wife

F is position of friend

t is threatened future

d the default future represents a temptation for the drunk, assisted by his friend, to defect from **P**

# indicates preference ranking (1 is most preferred)

Table 4. The Drunk, His Wife, and His Friend.

The friend's position, F, is that the husband should be allowed to drink (the friend likes someone to drink with) and the wife should let him have the money to do so (rather than forcing him to rely on the friend). The husband (the drunk) still swears to his wife that he will quit; thus the drunk and the wife share the position P. The drunk accepts P because he fears the threatened future t; however, he knows and she knows that he will be tempted to get his friend to assist him in defecting from position P: both the friend and the drunk prefer column d over position P. They can get to d from position P just by changing their own joint card selection, assuming the wife does not change hers. Note that the wife may not be able to detect a move from P to d. Because d is the default future (i.e., the future they are presently in, and will continue to be in if present policies continue to be pursued), the wife suspects that despite his promises, he will secretly carry on drinking, supplied by his friend.

Eliminating the Cooperation Dilemma

A character that faces a cooperation dilemma must, if it wants itself and others to genuinely accept its position, make them believe an incredible promise. How can the character do this?

The character may abandon its position. For example, the ethnic militia may openly say it does not agree to free movement, or the drunk may tell his wife he will not give up drinking. In each case (although not in all cases), this lands the character in other dilemmas; however, it eliminates the cooperation dilemma.

If a character does not abandon its position, it must try to make its promise credible.

What emotion then moves the character? In general, positive emotion, arising from the urge to reassure, will cause the character to feel and show goodwill, friendship, or love (the appropriate term depends on the types of characters and relationships involved). If the character seems sulky or reluctant or aggressive, its promises will not easily be believed.

Positive rationalizations go with positive emotion. The husband, in an attempt to be convincing, may explain how and why he has decided to change his life. The militia may give reasons why it has changed its mind and decided to agree with free movement. Unless reasons are given, the character's conversion will be unconvincing.

Emotion and reason are to some extent substitutes. Strong emotion without adequate reasons may convince others that the character can be trusted, temporarily at least. They may think, "He feels this strongly enough to go against his own interests or desires" (irrationally, in the game-theoretic sense of rationality). On the other hand, strong reasons may convince others of the character's trustworthiness, even if the reasons are accompanied by the wrong emotional signals. Emotion and reason are most effective when they accompany and support each other. Then the character's emotion shows that its preferences are in the process of changing (so that in the game-theoretic sense, the character must be temporarily irrational, while moving between preferences); the character's reasons give cause to believe the change will hold.

Finally, a character may take irreversible actions to eliminate the cooperation dilemma by removing its own temptation to defect. For example, the drunk may tell his accommodating friend to leave the house. The commander of the ethnic militia may make public broadcasts threatening discipline against those who set up roadblocks, thereby irreversibly tying his reputation as a leader to their removal and making himself not prefer *not* to take action against them.

Deceit, Disbelief, and Rational Arguments in the Common Interest

Is a character that projects emotions and gives reasons in the manner suggested above deceitful? Possibly, but not necessarily.

Irreversible actions actually may change preferences, as in the examples given. In addition, attitudes, values, preferences, and beliefs can change for emotional-rational reasons. We propose that it is just these forces of emotion and rationalization that cause genuine change. In convincing others, you may convince yourself. Indeed, if genuine change were impossible, deceit would be impossible. It generally is not possible to make others believe the impossible.

However, it is important that many of the above methods of eliminating the cooperation dilemma may indeed be carried out with deceit (i.e., not carried out in actuality, only in pretense). Deception means that the dilemma is eliminated from the characters' common reference frame (their common view of the situation, needed for communication). That is all that dilemma elimination requires. Any deceit, if successful, will not be discovered, at least until the Implementation phase, as shown in figure 2. Whether it is discovered then, later, or ever depends on characters' information at later stages.

Now any situation in which deceit is profitable is one where disbelief is appropriate. Any of the dilemma-elimination methods described above may be counter-productive and induce disbelief: "They want me to believe this; maybe I shouldn't." Hamlet says on such an occasion: "Methinks the lady doth protest too much."

We derive a theorem: "No one should ever believe anyone, because if you tell me something I can deduce that you want me to believe it, which gives me a reason not to, since presumably you would want me to believe it whether it were true or not (since in either case you would succeed in making it part of our common reference frame)."

To rescue us from this conceptual abyss, recall that it is the characteristic of reason and evidence to compel belief. They have this characteristic because they make deceit difficult or impossible. Of course, "difficult" is not "impossible." Reason and evidence may require exhaustive scrutiny to compel belief. However, it is their essential characteristic that they can do so because they can uncover deceit.

This, then, is why reason and evidence are needed to back up the more primitive method of emotion in attaining credibility. They are needed to overcome disbelief. We may speculate that humans developed rational argument in the course of evolution simply because they developed the capacity to deceive. We may reason as follows: After humans began to find an evolutionary niche in their enhanced ability to cooperate, deception became an advantageous strategy for an individual or subgroup by enabling them to share in the payoff from cooperation without paying their dues; in the presence of deception, disbelief became advantageous; and given disbelief, the ability to compel belief by rational argument became advantageous in turn.

Rational argument requires an ultimate framework of the common interest within which to operate, because it is always possible to go behind any argument or evidence and ask, "Why? How? Where from?" For this basic reason, arguments must in the end be founded on common values to be convincing. Common interests will, however, always exist when there is need to make a promise credible because such a need implies a common preference for having the promise believed and kept (as in column P) rather than not believed (as in column *t*). This common interest, generalized, can be made a foundation for common values and hence for acceptable arguments.

The ethnic militia, if it truly wants to convince the commander of its sincerity, may draw upon a range of common-interest arguments and a fund of possible common values in favor of freedom of movement and against setting up roadblocks. Similarly the drunk may draw on the values and interests he and his wife have in common in their marriage, all of which will be sacrificed if he continues to drink. In each case, the common-interest arguments made may need to be backed up by evidence of the value change they argue for.

In summary, we can derive from characters' needs to make their promises credible part of a process by which they build up or maintain a conception of their common interest and so become an effective single supercharacter in larger superconfrontations. From other dilemmas, we will derive other parts of this process.

Friction

Whether it is easy or hard to change attitudes, beliefs, values, and preferences depends on the strength of what must be changed. It is easy for me to prefer that you precede me through a door, although note that I feel an access of goodwill toward you even as I make this slight preference change. It would be hard for me to prefer to die to save you.

The many factors that make it hard to change a common reference frame are called *friction.* They involve many things. In general, friction arises when values are hard to change and evidence hard to escape.

In this general sense, however, values may be more changeable than the word suggests, and involve no more than trivial preferences or marginally different ways of applying deeply held value systems. For example, an officer who values discipline may make many small decisions in the course of applying it; these represent adjustments to what he means by this overarching value, and thus they are mini-value changes.

Why Do People Keep Their Promises?

Why is it we often can rely on people to keep a promise to do things that they would prefer not to do if they had not promised? Often it is because the friction they find in changing their preferences between futures is slight enough to be overcome by the emotional urge to come to an agreement and prefer to do as promised. They feel an emotional, preference-changing, dilemma-eliminating urge, however slight. To see this, recall how you feel when promising to do something you would prefer not to. You tend to feel and project goodwill, provided you mean to keep the promise. That projection of emotion tells others you can be trusted. If it is not there, people would find it disturbing (provided a preference change is indeed required; provided you are promising something you would have preferred not to do, rather than just notifying people of something you would want to do anyway). The same projection of positive emotion occurs when people change their beliefs to reach an agreement (dilemma elimination by belief-change).

In a further twist, strong, principled believers in keeping their promises may show their trustworthiness by projecting no emotion, signaling that having made a promise, their principles infallibly kick in to make them want to keep it. Such people lay themselves open to being misunderstood by strangers.

When stakes are high, such as in matters of peace and war, friction is too great for such social mechanisms. Merely promising something will not change preferences toward performing it.

The militia in table 2 wants to convince the commander of its sincerity, to keep him from forcibly freeing up the road network, but there is too much friction involved (the militia wants to keep roadblocks in place as a source of revenue, to keep out returning refugees, and to control movements of groups that might threaten the militia) for the militia to change its preferences and genuinely prefer free movement. The militia cannot help being insincere, despite wishing to convince the commander of its sincerity.

Eliminating the Cooperation Dilemma by Changing Position

So far in our discussion of ways of eliminating the cooperation dilemma, we have assumed that a player will maintain the aim of having its position accepted.

A player's proclamation that it aims to achieve its position may be deceptive. We have allowed for this. The drunk or the ethnic militia secretly may intend to defect, after having obtained acceptance of its position, but a character practicing deception still wishes to have its position accepted: it needs this to be able to defect from it. In light of his wife's ultimatum, "Stop or I'll leave you," the drunk cannot carry on drinking until he has made his wife believe he genuinely intends to stop. In light of the commander's ultimatum, "Stop or we'll use force," the militia commander cannot carry on obstructing roads without first convincing the commander he intends to stop.

A dishonest negotiator is still a negotiator that aims to have its position accepted.

By contrast, another way to escape the cooperation dilemma is to actually change one's position (i.e., to switch to a position that does not suffer from that particular dilemma in that particular way, although it may suffer from other dilemmas and even from another cooperation dilemma).

What emotions accompany the abandonment of a position? If a player has been determined to maintain its position, sorrow and despair are feelings that help it to withdraw emotionally from the objectives it has pursued, reassess the situation, and pursue less ambitious aims. These emotions generally accompany a committed player's change of position.

In table 4, the husband would feel sorrow and despair if his inability to stop drinking made him decide to give up his wife (i.e., if he decided that the only way to solve his cooperation dilemma at P was to shift position by accepting the threatened future *t,* consisting of his wife leaving him while he continues to drink with his friend).

Alternatively, a player may welcome a change of position. This is likely to be so with a player facing a cooperation dilemma: "Because I got myself into this dilemma by accepting a position that is inferior to at least one other possible position (the one I am tempted to move to), I can probably think of other positions I would prefer to go to, if I could get other players to accept them. For example, there may be a position I have relinquished unwillingly, under threat of sanctions, but would like to go back to it if I can persuade others to agree."

A player who for such reasons would welcome a change of position may seize upon the dilemmas that beset its present position as an opportunity to persuade others that it might be a good idea to move away from it. We can suppose, for example, that the ethnic militia in table 2 and the drunk in table 4 each have accepted position P solely to escape *t*. This being so, they may use their own cooperation dilemma at P as an argument for abandoning or modifying this position. The drunk may say, "I know I agreed to stop drinking, but you know how hard it will be for me. Would you really be able to trust me? Maybe we should think of some compromise arrangement I would actually be able to live with." The militia commander may say, "Agreed. Free movement. But it will be tough to get my people to agree. I don't know if I can do it. Is there a compromise position it would be easier for me to sustain?"

The emotions that accompany attempts to change position in the latter case (when the aim is to move to a preferred position, as distinct from one that is accepted only because it is realistic) are, like despair and sorrow, emotions of withdrawal. The emotional tone here is skeptical, detached, and deflationary, rather than despairing, because the character needs to deflate emotions and puncture arguments that suggest commitment to the present position by pointing to other ways of solving its dilemmas.

If the wife or UN commander beams forth positive emotion and constructive arguments as to how the drunk may stop drinking or the militia commander may free up the roads, the drunk or militia commander, if they want to argue for a move to another position, will try to deflate the emotion and debunk the arguments by being skeptical, realistic, and emotionally uninvolved with the problem. In this way, they hope to initiate acceptance of another, more acceptable position.

THE TRUST DILEMMA

I face a trust dilemma when you are tempted to defect from my position.

In card-table terms, a trust dilemma confronts me when a group of one or more characters (not including me) can, by changing just their own card selections, move from my position to another future they all like just as well.

Note that when we all share a common position and I have a trust dilemma, it must constitute a cooperation dilemma for someone else.

In table 2, the commander has a trust dilemma because he cannot trust the militia to allow free movement. In table 4, the wife's trust dilemma is that she cannot trust her husband not to continue drinking alcohol supplied by his friend.

Is this the same dilemma stated twice? It is not, because my inability to trust you is my dilemma; your inability to be trustworthy is yours. There is a difference between not being able to trust myself (thus being potentially untrustworthy to others) and being unable to trust others. Consider, for example, the case when the ethnic militia in table 2 wants to escape its cooperation dilemma by changing position, while the UN commander wants to get rid of his trust dilemma by persuading the ethnic militia to change its attitudes or incentives. His trust dilemma is the militia's cooperation dilemma. Their perspectives are different.

When there is no common position, my trust dilemma does not need to be someone else's cooperation dilemma. In table 3, the local commander has a trust dilemma because he is asking the local militia to agree to something he cannot trust them to implement.

For example, if the militia agrees to remove the roadblock (column C), it can move from C to its own position, M, simply by taking back the "Remove roadblock" card (the militia prefers M to C). The militia can move unilaterally from the local commander's position to a future it prefers. This gives the local commander a trust dilemma. The lesson he can draw from it is this: If the militia agrees, he will need to make sure it does remove the roadblock, and does not put it back when he goes away.

The same dilemma faced the United States when President Richard Nixon and Secretary of State Henry Kissinger attempted to negotiate with North Vietnam to end U.S. involvement there. Put simply, the U.S. position was: "We'll leave Vietnam if you agree not to invade the South." North Vietnam replied: "How can we agree to such a thing? Would you believe us if we did?" The U.S. response was: "Agree, damn you, or we'll bomb you until you do." And the United States did just that; it dropped more bombs on Hanoi than had been dropped in the whole of World War II. The North Vietnamese finally said: "Okay, we agree." The U.S. response was: "We don't believe you." How could we? Still, the United States left. North Vietnam then invaded the South.

Clearly this account simplifies the negotiations; nevertheless, it captures the essence of the United States' trust dilemma. In a moment of truth, negotiators themselves simplify the issues to make sure they have a common reference frame.

Eliminating the Trust Dilemma

As with the cooperation dilemma, I can eliminate my trust dilemma by giving up my position. For example, the local commander can cease to require the roadblock to be removed. The wife can agree to let the husband drink.

The emotions that go with giving up a position one is committed to are generally negative toward oneself (e.g., despair, sorrow). Accompanying rationalizations consist of finding reasons why the new position is not so bad, after all, and may even be better than the one abandoned. As with the cooperation dilemma, it is also possible that the position being moved to is already preferred, in which case the trust dilemma may be welcomed as an argument in favor of changing position.

To maintain my position against a trust dilemma, I must find ways to eliminate another's temptation to defect from it. The required emotional tone is then positive and constructive, showing goodwill, cooperativeness, and empathy with the other. This is true even if I impose an automatic sanction to eliminate the other's temptation.

Suppose the local commander lets the local militia know that, after the roadblock has been taken down, a reconnaissance satellite will spot it if the roadblock goes up again, and rocket fire automatically will be called down. Despite the fact that violence is being threatened, a positive, friendly tone will convey this information most effectively, because its object is to elicit cooperation with the commander's position. The commander's tone should imply that violence will be unnecessary because the militia will cooperate. Violence is there to stabilize the agreement and provide a reason for the parties to trust each other.

Rationalizations accompanying the projection of such positive emotion also should be positive, aimed at constructing a common position on the basis of common interests. For example, the local commander, even while giving notice of his automatic sanction, should adduce all kinds of common-interest arguments in favor of freedom of movement, such as rebuilding the economy and bringing the country forward to join the community of advanced nations. The aim is to bring about willing compliance with his position. He should assume this will be forthcoming and welcome it in a spirit of goodwill.

The wife might let her husband know that she can always spot his drinking and will leave without further warning if he continues one more time. This is automatic retaliation. At the same time, she should build on their love for each other and their common interest in the marriage succeeding provided he gives up drinking.

Deceit, Disbelief, and Positive Argument

The local commander's threatened satellite surveillance and automatic calling of fire may not exist or be as efficient as he suggests; the wife may not be able to spot her husband's drinking as infallibly as she says; thus, these communications may be more or less deceitful. Because this is so, the other side may be more or less disbelieving. The fact that one side has reason to produce belief, even if unjustified, makes disbelief rational for the other side.

To combat disbelief, concrete evidence may be produced that automatic retaliation is a reality. Alongside or instead of this, positive arguments in favor of the position advocated are a way to deflect disbelief. A bald sanction invites disbelief and hence resistance, leading to thoughts of how to get around it. Positive reasoning on the basis of common interests directs attention away from such reactions and toward seeing the threat of automatic retaliation as an excuse for giving in to what, judged by such reasoning, seems preferred on its own merits.

Thus rational common-interest arguments, positive emotion, and evidence of automatic retaliation are, in a way we have seen before, both possible substitutes for each other and strongest when used together. If my system for automatic retaliation is objectively unconvincing, I may compensate for this by good common-interest arguments, presented sympathetically. At one extreme, I may be able to do without retaliation altogether and change another's mind merely by common-interest arguments in favor of my position; vice versa, I may compensate for weak arguments presented badly by strong evidence of mechanisms for retaliation.

Modifications of Position

Your arguments will be most effective if you take account of others' attitudes and beliefs and pay attention to their objections, incorporating them into your position if possible.

In this way, their objections can be used to modify your position itself. Assuming that your aim is to maintain your position, not to change it, you may nevertheless find that one way to safeguard the essential features of a position may be to modify it to make it more acceptable to others. This works as follows. Building upon common interests naturally can lead to the addition of cards that make our position more favorable to the other, eliminating the dilemma by making our position better for them than the temptation to defect. Thus the wife or local commander might promise the husband or local militia something they have been wanting as a reward for going on the wagon or removing the roadblock.

Such inducements will be more credible if stated in terms of the common interest rather than in the interests of those you are inducing, because common-interest arguments give them reason to suppose you will want to carry out the promise you are making. They also will be more credible if first suggested by themselves, because this makes the inducements seem less likely to be a trick or trap.

History cannot be rewritten; nevertheless, it is interesting to speculate how the United States might possibly have eliminated its trust dilemma with North Vietnam. Such speculation might include creating new cards to put in characters' hands: automatic retaliation (U.S. forces stationed nearby and conducting reconnaissance), positive arguments (suggestion of a U.S.–Vietnamese alliance against Russia and China), or modification of position (a federal Vietnam with South Vietnamese rulers maintained in power through elections).

Friction

Suppose, as the UN commander, you want to eliminate a trust dilemma without substantially changing your position. You proceed to assess evidence and arguments produced by the other side (the ethnic militia) to prove it does not intend to defect. As you do so, it will be inherently harder to eliminate your dilemma (a) the more mistrustful you are and (b) the greater your estimation of the cooperation dilemma the other would face if it were to accept your position (or does face, if it has already accepted it).

These are distinct sources of friction. The first, mistrustfulness, should be avoided. A highly successful businessman said, "I trust everyone until they give me reason not to." He was explaining how to make money out of a tendency to be over-mistrustful (arising from the fact that in modern society we frequently deal with strangers).

Naïve, mistaken acceptance of others' assurances is best avoided, not by mistrustfulness, but by realistically assessing the reasons for their cooperation dilemma and helping to counter them in a positive spirit of cooperation, even if, as noted, you are asking them to cooperate with damaging acts of automatic retaliation such as jets bombing roadblocks.

The bad effects of mistrustfulness are that, instead of encouraging others to genuinely adhere to your position, it invites them to treat it as a position that will have to be abandoned. By projecting a skeptical, detached attitude and treating your trust dilemma (which is their putative cooperation dilemma) as something hard to overcome, you give the impression that you consider your own position untenable.

Willing Changes of Position

I actually may prefer to move, with others, to another position. If this other position does not suffer from the trust dilemma I face at my present position, I may welcome my trust dilemma because it gives me an argument for switching.

Suppose the wife actually wants to leave her husband, the drunk. Her position, that she will stay if he will stop drinking, is one she would be happy to relinquish. She will then welcome her trust dilemma as giving her a reason to leave him. Whatever he says or does she will not want to trust him. Her mood will be skeptical and uninvolved so as to deflate any emotion and argument he may use to try to persuade her.

Suppose the local commander (belonging to a coalition nation that lacks high standards of military training) has been bribed to let the roadblock stay. His position, that it be moved, is then one he wants to give up. He accepts with detachment his trust dilemma, the fact that he cannot trust the local militia not to put back the roadblock as soon as his back is turned. He deduces from it that there is no point trying to get the roadblock moved.

SUMMARY OF CHAPTER 3

To win a confrontation by getting others to accept his position, a commander needs to understand how changes come about at the Climax phase. Game-theoretic dilemmas are the key. They face players that try to pursue fixed objectives within a fixed frame, causing them to feel emotion and to seek reasons to change their preferences and the frame. The metaphor of dramatic interaction seems better than game-playing for describing such processes of transformation; hence confrontation analysis is based on drama theory, an extension of game theory. Through trying to influence each other by arguments in the common interest, dramatic characters tend to build up the attitudes and interests of a supercharacter.

Six game-theoretic dilemmas pressure characters at a moment of truth: the dilemmas of cooperation, trust, deterrence, inducement, threat, and positioning. The cooperation dilemma faces a character that belongs to a group that can do better by defecting from it (the character's) own position. It is the dilemma of a drunk trying to persuade his wife he means to stop drinking after she has threatened to leave him if he does not. Eliminating this dilemma without greatly changing one's position means making an incredible promise credible. It requires positive emotions and rationalizations as to why you do not now prefer to defect. Such protestations may be deceitful, inviting disbelief, which finally can be overcome only by sound reasoning and evidence. Friction to be overcome involves hard-to-change values and beliefs. Some friction is relatively slight, like that involved in keeping easily-kept promises. Promises involving high stakes need to be guaranteed with convincing reasons, conversions, or sanctions.

One way of overcoming a dilemma is to change your position to one that does not suffer from this dilemma (though it may suffer from others). If you are committed to your position, emotions of sorrow and despair may help you to rationalize the switch to another position. You may already prefer the other position, in which case you may use the dilemmas of your present position as arguments in favor of a move.

The trust dilemma faces a character when a group not including the character is tempted to defect from its (the character's) position. If it does not wish to give up its position, it must eliminate the trust dilemma by eliminating others' temptation to defect, either by arranging sanctions against it or by reasoned persuasion and modification of its position to others' benefit. The required emotional tone is positive, even when arranging punishing sanctions. Deceit is a temptation and disbelief problem. Mistrustfulness is counter-productive because it is negative in tone.

CHAPTER 4

MORE PRESSURE: DILEMMAS OF DETERRENCE, INDUCEMENT, THREAT, AND POSITIONING

THE DETERRENCE DILEMMA

I face a deterrence dilemma in relation to a character that is under no pressure to accept my position. It feels no pressure because the threatened future does not deter it from rejecting my position; rather, it encourages it to.

In card-table terms, I have a deterrence dilemma in relation to a character opposed to my position that prefers the threatened future to it (my position).

Unlike the cooperation and trust dilemmas, this dilemma arises only when parties take different positions. There are conflicting positions in the local commander's confrontation (see table 3), but not in the problem faced by his superior (see table 2). In table 4, the drunk and his wife share a position, but it conflicts with that of the friend.

The high-level, general confrontation in the former Yugoslavia between the Serbs and the West placed the West in a deterrence dilemma for several years.

The West wanted the Serbs to desist from ethnic cleansing. If the Serbs refused, the West threatened to impose economic sanctions, and actually carried out this threat. The economic sanctions were no deterrent because the Serbs as a whole (taking into consideration the dynamics of the interaction between the Bosnian Serbs and the Serbian government) preferred the threatened future, "Ethnic cleansing with economic sanctions," to "No cleansing, no sanctions," which was the West's position.

Consider the local commander's confrontation in table 3. The local militia has not yet taken a definite fallback position; therefore, what is represented is a point in the Buildup phase of figure 2 at which both parties, the local commander and local militia, have taken up positive positions. The commander wants the roadblock removed; the militia wants it to stay. Also, the local commander has effectively stated the following fallback position: "If you don't remove the roadblock, I'll forcibly remove it. If you then fire on my troops, I'll call in air support." The local militia has not yet stated a fallback position. Whether a deterrence dilemma now follows depends on the fallback position the local militia chooses. A refusal to remove the roadblock (without a threat to fire on Allied troops) puts the militia in a deterrence dilemma because the local commander prefers the threatened future t (forcible removal of roadblocks) to M, the militia's position; therefore, this would not pressure the commander to accept M.

Eliminating the Deterrence Dilemma

The deterrence dilemma, like other dilemmas, may be eliminated by giving up one's position. For example, the West might have agreed to let the Serbs have their way, leading to most of Bosnia being divided up between Serbs and Croats, with large-scale genocide of Muslims. Such a withdrawal policy was advocated by many in the West. The local militia could solve its deterrence dilemma, if it had one, by agreeing to remove the roadblock.

Suppose, however, that a character is determined to retain much of its position. An alternative way of eliminating the deterrence dilemma is to escalate to a higher level of retaliation to make the threatened future worse for the other party. This generally involves thinking up, or bringing onto the agenda, new cards with which to punish them. This process of thinking up and making credible new punishment cards is driven by negative emotions such as anger and indignation. It is rationalized by demonizing the party to be deterred; that is, seeing them as wicked and evil, so that it becomes right and necessary to consider extreme reprisals against them.

In the overall confrontation between the West and the Serbs, Western public opinion demonized the Serbs (while ignoring, or paying less than proportionate attention to atrocities committed against them by Croats and Muslims) until it was possible for serious military intervention against them to become a credible part of the threatened future, eliminating our deterrence dilemma.

The West speedily demonized Saddam Hussein (for example, relative to President Assad of Syria and others) when he invaded Kuwait, and it became necessary to get rid of a deterrence dilemma by threatening military force against him.

In table 3, the militia can get rid of its deterrence dilemma by adopting the threatened future t', rather than t; that is, by threatening to fire on Allied troops. Though this invites retaliation from the air (this being part of the commander's fallback position), it cures their deterrence dilemma because the danger of allied casualties means the commander prefers M (the militia's position) to t'. In making this threat, they are likely to be driven by feelings of resentment and anger.

Escalation and De-escalation—Conciliation as a Response to the Deterrence Dilemma

Thinking up new, more damaging threats is a negative, escalatory way of resolving the deterrence dilemma. It may be necessary, but it is risky. It runs the risk of triggering a process of mutual demonization and mutual escalation, in which each party responds to threat escalation by further escalation, driven at each stage by increasing feelings of hatred and anger rationalized by viewing the other as more and more evil.

This risk is greatest when characters have roughly equal ability to escalate their threats. Clausewitz (1968; 1st edition 1832) generally assumes war between equally matched states. Accordingly, he sees mutual escalation as an inherent tendency of war. By definition, he thinks war implies "a sort of reciprocal action, which must logically lead to an extreme" in which "even the most civilized nations may burn with passionate hatred of each other" (p. 103). Clausewitz, however, does not see why such passions are aroused; that is, he does not see their drama-theoretic function.

As we saw in chapter 1, war today is not generally Clausewitzian. One side (the United States and its allies) normally has superior capacity to escalate its threats. This imbalance may diminish the risk of mutual escalation, often enabling the allied side to cure its deterrence dilemmas by one-sided escalation, provided it is done in the right manner.

Consider a positive, conciliatory way of resolving the dilemma. This is to add new cards that improve (sweeten) our position in the estimation of the threatened party. In this way the threatened future becomes worse for them than our position, not because we have made it worse, but because we have made our position better. The emotion that goes with this method is the positive one of reaching out and sympathizing with the other's needs. Concomitant rationalizations are designed to accomplish the following:

- Prove to ourselves that others have genuine needs that deserve to be met

- Establish our position, as now modified and improved, as a win-win outcome that they ought to accept.

Following are some examples:

- The local militia, rather than escalate by threatening to fire on the local commander's troops, might try to persuade him to accept the roadblock as a means of keeping peace in the area. "We'll always let UN troops through," the militia might say. "We can use the roadblock as a way of helping you accomplish your mission."

• In February 1994, U.S. envoys reportedly offered President Tudjman of Croatia a deal: stop fighting the Muslims of Croatia and we will let you take back the Krajina region from the Serbs (Silber and Little, 1996). The ensuing de facto alliance between Croats and Muslims led to a succession of setbacks for the Serbs that persuaded the Bosnian Serbs to accept the Dayton agreement. Thus, the United States shifted its position in favor of the Croats to persuade them to stop fighting the Muslims, and join them in fighting the Serbs.

• The 1978 Camp David agreements between Israel and Egypt brokered by U.S. President Jimmy Carter were made possible by the United States sweetening its position for both sides by promising each of them large sums in aid.

By adding new cards that sweeten our position, we do run the risk of creating a cooperation dilemma for ourselves. Others may suspect we will not keep our promise. If this occurs, we must use previously described positive-emotion methods for overcoming a cooperation dilemma by giving reasons why we should be trusted. Our cooperation dilemma is their putative trust dilemma.

Note that conciliation and trust-creation both involve positive emotion. Following are the differences:

- Conciliation consists of modifying our position to make it more attractive to others, thereby eliminating a deterrence dilemma

- Trust creation eliminates a cooperation dilemma by giving reasons why we should carry out our promise (i.e., why we should prefer our own position to a possible temptation to defect from it).

Conciliation needs to emphasize how our position benefits them, trust creation how it benefits us. Common-interest arguments have the virtue of emphasizing both points at once.

Conciliation (i.e., enhancing our position from the viewpoint of the other side) also has been discussed as a method of getting rid of the trust dilemma. The difference is that removal of the trust dilemma addresses a situation in which both parties share the same position, as with the UN theater commander in negotiations with the ethnic militia. It assumes either that this is the situation or that the trust dilemma is the main objection to making it so (as with U.S. negotiations with North Vietnam). Removal of the deterrence dilemma, by contrast, tries to rectify a situation in which the other side is refusing to accept our position because they prefer the threatened future. It is a mistake to confuse these situations.

U.S. policy in negotiating with North Vietnam, if our analysis is correct, made just this mistake. It dealt with a trust dilemma (i.e., "How can we trust them not to invade the South after we leave?") using methods appropriate to a deterrence dilemma (bomb them until they agree not to invade the south). These methods brought the North Vietnamese to the conference table, but left our trust dilemma unresolved.

Conciliation Combined with Escalation

Conciliation as a method for overcoming the deterrence dilemma has a disadvantage; it may mean we must make distasteful concessions. Conciliation became particularly discredited by British Prime Minister Chamberlain's attempts to conciliate Hitler in the Munich crisis of 1938; conciliation was famously used then as an alternative to escalation, because Britain and France were unwilling to threaten war.

Conciliation may be effectively combined with escalation in a manner that makes large concessions unnecessary. Pure escalation has a negative emotional tone and rationalization; it tends to rationalize a preference for anything that is against the others' interests. Consequently others will fear that their interests will not be safeguarded in any settlement they might discuss with us. (Recall that the exact details of any settlement are discussed in the Resolution phase following overall agreement.) Consequently, rather than discuss a settlement at this stage, they will look for possibilities of counter-retaliation to lay the basis for a more balanced solution. In this way, pure escalation encourages counter-escalation, if possible.

Combining escalation with conciliation in a "tough cop, tender cop" routine, gives the impression that while determined to punish refusal to settle, we are willing to be sympathetic to others' interests if they do settle. This impression may count for more than any concrete concessions you offer, allowing actual concessions to be relatively slight. The point is that others need negative feelings and rationalizations directed against us to take the path of escalation. Our concern for their interests undermines such negative feelings and makes it harder for them to justify escalation. In terms of the internal confrontations that determine their policies, conciliation gives the doves within them arguments against the escalation-favoring hawks.

Friction

The deterrence-dilemma friction consists of any inherent difficulty in making others believe that the threatened future is worse for them than your position. This friction may be great or small.

For example, in 1995 at Wright-Patterson Air Force base in Dayton, OH, Holbrooke took Milosevic and Bulatovic (the Montenegrin President) into the Nintendo room, a map center equipped with computers that allowed the user to overfly the Bosnian terrain. At this point the positions of the Serbs and the United States were close, differing only in regard to whether a certain area should be under Serb control. Showing him the area in dispute, Holbrooke said, "Am I seeing right? There's nothing there. Just mountains. No houses, no villages." Bulatovic said, "That's right, but this is Bosnia." Holbrooke responded, "Look at what you're fighting for. There is nothing there." He was producing reason and evidence in favor of the U.S. position by showing the Bosnians how close it was to their own, and hence how much they should prefer it to the threatened future. (See Silber and Little, 1996, p. 373.)

When at the Athens meeting in 1993 Cyrus Vance told the Bosnian Serbs "the U.S. Air Force is all prepared to turn Bosnia and Serbia into a wasteland," he was giving reasons why they should consider the threatened future to be worse for them than the U.S. position.

Overcoming this type of friction is a matter of working on others' beliefs and values to affect their comparison between our position and what they can hope for if they reject it. Making our position or fallback position credible, given the extra cards we may have added to them, is a separate matter of solving either a cooperation dilemma or one of the three remaining dilemmas examined below. It is not a matter of the deterrence dilemma as such.

Subcharacters Aiming to Change Your Position

We have said you can eliminate a deterrence dilemma in a conciliatory manner by changing your position in the sense of giving up something; however, such a change of position may be just what certain subcharacters belonging to your organization prefer. They may use your deterrence dilemma as an argument in favor of the position change that is their objective.

- For opinion holders in the West who preferred not to oppose the ambitions of Hitler, the Soviet Union, China, North Vietnam, the Serbs, and Saddam Hussein, the West's deterrence dilemma ("How can we deter them?") was not a problem but an opportunity. They used the answer, "We can't," as a reason to stop opposing.

- Similarly, for opinion holders in the 1980s in the Soviet Union who wanted the Soviets to move toward the West's position, the Soviet deterrence dilemma ("How can we match the Reagan arms buildup to negotiate from strength?") was not a problem. They too would have argued, "We can't. That's why we should give in."

When internal confrontations take place to determine the policies of large characters such as the West or the Soviet Union, a subcharacter's true objectives may be revealed by its attitude toward dilemmas. One that really wants to achieve the organization's stated position will show negative emotion in face of a deterrence dilemma. One whose true aim is to change that position will be more detached and objective.

Detachment and objectivity go with a tendency to regard the frame as fixed, encouraging positional change because if the frame is fixed, then the only way to escape from a dilemma is to change position.

THE INDUCEMENT DILEMMA

The inducement dilemma is the other side of the deterrence dilemma: In eliminating my deterrence dilemma, I give you an inducement dilemma.

In card-table terms, a character has an inducement dilemma if another's position (though different from its own) is as good for it as the threatened future.

It follows that by successfully solving a deterrence dilemma I give someone else an inducement dilemma. However, I may give myself one as well.

When the West, in its overall confrontation with the Serbs, finally started threatening armed intervention, it replaced its deterrence dilemma with an inducement dilemma. It preferred the Serbs' position ("We continue ethnic cleansing") to implementing its new fallback position (i.e., attacking the Serbs). This became obvious to the Serbs as the West, in one subconfrontation after another, failed to follow up on its threats. In terms of the five possible endings in figure 2, each confrontation ended with the implementation of a threatened future flunked by the West.

After Munich, as Hitler's ambitions proved to be unappeased, Britain and France started to rearm, thereby making war part of their threatened fallback position. In this way they may have eliminated their deterrence dilemma (Hitler might have preferred to back off rather than fight) but replaced it with an inducement dilemma (Hitler's disbelief in their willingness to fight).

These examples illustrate that while my deterrence and inducement dilemmas may be similar in their effects (in that both may allow my opponents to have their way undeterred by me), they are different in their causes and need to be tackled differently. My deterrence dilemma is a matter of my opponents' preferences. I must ensure that they prefer to accept my position rather than provoke me into taking up my fallback position. Having ensured this, I may still face an inducement dilemma, which is a matter of my own preferences. The question here is, do I prefer to give in to them rather than implement the threatened future? If so, I have an inducement dilemma to overcome to make my deterrence fully credible.

Avoiding Escalation by Accepting the Inducement Dilemma

By overcoming my inducement dilemma, I give the other side a deterrence dilemma. This forces them to choose between giving in or overcoming that dilemma. The latter choice gives me another inducement dilemma to overcome, and so on. For both sides to try to overcome both dilemmas means embarking on a cycle of escalation.

Therefore the question arises: Is it possible to negotiate in a situation when both sides have an inducement dilemma and neither has a deterrence dilemma? The answer is yes.

Under the nuclear deterrence regime of the Cold War, both sides were in this position. Each must have preferred, at each confrontation, to give in to the other rather than start a nuclear war. Because this situation was clearly symmetrical (balance of terror), neither side simply gave in; they negotiated.

Hitler and Stalin negotiated the Nazi–Soviet pact in a similar way. Each preferred a range of positions to a Russo–German war because for each of them those positions meant a license to gobble up their neighbors, their military forces being the only ones in Europe that presented a serious threat. (Each knew, of course, that after this gobbling-up process had reached its limits, all bets would be off.)

Often in such negotiations the threatened future is tactfully left unmentioned by the parties. They negotiate positively, advancing their own positions and attacking others' on the grounds of each position's contribution to the common interest. Because all positions being discussed are preferred to the threatened future by all parties, continually comparing them with that threatened future is unnecessary.

Balanced negotiations of this kind, in which each side, rather than suffer a breakdown, would prefer to accept a range of positions, are the norm in civilized relations (e.g., most trade negotiations). Such negotiation situations are not stable. They contain dilemmas, causing them to move; however, they are prevented from escalating and kept within stable bounds by directing the destabilizing effect of the inducement dilemma toward impelling characters to negotiate a single position. This is a way of eliminating the deterrence dilemma that is a non-escalatory alternative to preferring the threatened future to others' positions.

After they have agreed to a single position, the characters generally will face cooperation and trust dilemmas. Foreseeing such dilemmas will be one of the factors leading them to accept this or that position. Ideally the main factor leading them to agree will be passionately argued rational arguments in the common interest, even when the common interests involved are as crude as in the case of the Nazi–Soviet pact.

Other Ways of Eliminating the Inducement Dilemma

Should we then see balanced negotiations (i.e., initial acceptance of the inducement dilemma followed by rational, common-interest debate) leading to its elimination through convergence to a single position, as the norm for defense forces. Should we aim for it?

We probably should when considering relationships between internal actors (the subcharacters within an alliance whose confrontations determine its policies). We are then concerned with cooperative confrontations, as discussed in chapter 10. We probably should not when our defense forces face rebels against the New World Order. In such cases we typically find an extremist approach on the rebel side (i.e., a willingness to escalate threats to a point where they can dictate the outcome), and an asymmetry of potential power with the Allies, led by the United States, having superior escalatory capacity.

Opponents motivated by a philosophy requiring them to refuse any kind of compromise may reject the idea of negotiating on the basis of common interests. They may work themselves up into preferring a breakdown (i.e., the threatened future), no matter how dangerous, over any position acceptable to us. Our forces then may need to eliminate our deterrence dilemma by putting on the table cards that both punish and conciliate the other sufficiently to make them prefer at least one acceptable position to a breakdown. This means combining adequate deterrence with a conciliatory posture. It means carefully lining up our guns to point at their head, then saying, "Right, let's talk about your problems." It also means eliminating our inducement dilemma by preferring the breakdown (firing our guns) to acceptance of their extreme position.

Eliminating our inducement dilemma gives them a deterrence dilemma, forcing them to either modify their extreme position or escalate by making the breakdown still worse for us. We are trying to ensure the former response. To forestall the latter, we must eliminate our inducement dilemma so thoroughly as to prefer the breakdown to their extreme position, no matter how much they escalate. In general we have the capacity to do this because of our preponderance of military power, although it is true that new threats of chemical or biological attack on cities may enable terrorist groups to make the West back down.

U.S. and Allied preparations for the Gulf War, both military and psychological–political, exemplified thorough elimination of our inducement dilemma; it covered all possibilities of Iraqi retaliation and prepared us for the worst. The problem seems to be that we failed to cure our deterrence dilemma (i.e., Saddam Hussein still preferred war to leaving Kuwait). We may have assumed too much common interest between him and the Iraqi people, who were the ones that suffered from the war.

What, then, is the recipe for eliminating an inducement dilemma when the other side will not budge from an unacceptable position? Negative emotion and rationalizations are needed; however, they do not need to be directed entirely at the other side as they must be to eliminate the deterrence dilemma, where we need to think of cards that will hurt them. It is now our own preferences and underlying values that concern us. We need to think sufficiently poorly of the extremist position being rejected and reconcile ourselves sufficiently to the threatened future, to prefer the latter to the former. It can be a case of "hate the sin (the position being rejected), not the sinner (the characters taking that position)." A feeling of martyrdom ("By taking this position they are forcing us to accept the threatened future") may be the greatest negative feeling invoked against the other side.

The major argument leading a wide coalition of nations to support *Operation Desert Storm* was the unacceptability of letting Saddam Hussein get away with the annexation of Kuwait. Many reasons were given for this, taking account of different parties' interests. Hostility to Saddam Hussein was not important, except for the Allies who had to gird themselves to put the intervention card on the table. They were the ones who had needed to make an effort (unsuccessful though it turned out) to overcome the deterrence dilemma as well as the inducement dilemma.

Modifications or reappreciations of the threatened future that make it more acceptable to us, but not to the other side, also help eliminate our inducement dilemma.

For example, at the outbreak of World War I, as to some degree with any hostilities, there was patriotic pride and excitement at the prospect of using violence. This precisely fitted the bill, making the threatened future more attractive to us while encouraging creative ideas as to how to make it worse for them.

It might seem there also could be modifications of the other side's position, as distinct from reappraisals of it, to make it less attractive to us. There are suspected cases of this. For example, the suspicion that President Franklin Roosevelt permitted Pearl Harbor to happen or that in February 1994 Muslims planted the bomb that blew up their own people in Sarajevo. Here, covert direct intervention is suspected of having made a position worse in someone's eyes, thus overcoming an actual or potential inducement dilemma. Why must it be covert? It cannot be done openly because it is inconsistent to encourage something that we denounce (i.e., evil aspects of the other's position). Inconsistency is against reason, and reason is necessary to compel belief at a moment of truth. We admire inconsistency and confessed self-manipulation in our leisure moments, not at the moment of crisis in a confrontation.

Friction and Subcharacter Conflicts—the Inappropriateness of Cost-Benefit Analysis

The inducement dilemma may be hard to overcome, even for a character that can easily overcome the deterrence dilemma. It is relatively easy for the United States, leading the West, to threaten action against tyrants and murderers; it is harder to carry it out, particularly when U.S. citizens' lives may be lost. This can lead to a commander being given absurd mandates such as, "Threaten to do it; but don't do it without permission."

What is needed is the political will to take action if necessary, supported by an understanding that such willingness to carry out threats will mean that they will less often have to be carried out (provided the deterrence dilemma is first overcome, so that the threats are adequate). There may always be subcharacters within our character that want to shift position toward acceptance of an opposing position that is giving us an inducement dilemma. They will point to disadvantages of the threatened future as if our preference for it should be determined by cost-benefit analysis. But this tool is inappropriate when comparing the threatened future with a position we are rejecting, because we then have a paradoxical need to prefer to carry out our threats in order not to have to carry them out. Concepts such as honor, patriotism, solidarity, ethics, and adherence to principle meet this need. These emotional concepts have no role in economics, but they do have a vital role in resolving confrontations in which military action is a card that needs to be made credible.

THE THREAT DILEMMA

Often the threat dilemma coincides with the inducement dilemma, but it is conceptually and practically different. It occurs when I cannot be trusted; I cannot even trust myself to implement my part of the threatened future if we move into the conflict phase of table 2.

In card-table terms, I face a threat dilemma when I can move, just by changing my selection of cards, from the threatened future to another future I like just as well. Therefore, at the moment of truth, others will suspect that I am bluffing.

We saw in chapter 2 how the West flunked the implementation of Vance's threat "to turn Bosnia and Serbia into a wasteland" made in May 1993 to the Bosnian Serbs in Athens. The West faced a threat dilemma: it preferred doing nothing to launching a bombing campaign. It is unclear if the Serbs suspected them of bluffing, but the failure must have damaged Western credibility for the future. (See Silber and Little, 1996, p. 282.)

The dilemma often coincides with the inducement dilemma because the temptation I feel to refrain from implementing the threatened future may be a temptation to accept your position. However, I may have other temptations to defect to from the threatened future. In any case, the two dilemmas are different.

The threat dilemma is like the cooperation and trust dilemmas in that it arises from parties looking forward to what is likely to happen in the Implementation phase. It is a problem for me because others suspect I might not prefer to carry out my threat if and when the time comes to do so; therefore, they discount my threat as incredible.

The inducement dilemma, like the deterrence dilemma, is grounded in the tug-of-war that takes place at a climactic moment of truth. I am under pressure then and there to give in to your position, because to do so is as good for me or better than the threatened future; therefore, for purposes of the inducement-dilemma argument, the threatened future is assumed to be credible, not incredible. By contrast, my threat dilemma is all about questioning the credibility of the threatened future. You are disinclined to believe I will carry out my threat. If I walk out threatening to do so, you think I will either come back, having changed my mind, or do something else instead.

Getting Rid of the Dilemma

Despite this difference between the two dilemmas, many of the emotions and rationalizations we use to overcome an inducement dilemma work as well for a threat dilemma, for the simple reason that both can be overcome by raising our valuation of the threatened future. In this way we overcome the inducement dilemma by raising its valuation relative to another's position; we overcome the threat dilemma by raising it relative to any temptation we have to defect.

Helpful emotions are defiance, anger, indignation, and the martyred feeling of being forced into the threatened future by others' intransigence. Temptations may be downplayed, with reasoning as to why we would not want to do that anyway. The most general, all-purpose rationalizations center on elevating our preference for the threatened future. There are two kinds: first, evocation of principles such as honor, self-respect, integrity, and the need to keep commitments; second, evaluation of our fallback position as being instrumentally the best way, or at least a good way, to deal with the situation created by others resorting to their fallback positions.

The British continually have used both kinds of argument to justify their continuing fight against terrorism in Northern Ireland, despite the public's frequently-stated preference for "leaving them to fight it out." The same two kinds of argument were used in the United States to justify the continuing involvement in Vietnam and, more recently, the decision to fight Saddam Hussein.

Subcharacter Conflicts Over the Threat Dilemma

It is essential to note that friction to be overcome in eliminating the threat dilemma is felt in the Implementation phase itself. Its effect at the moment of truth or in the preparatory Conflict phase is derivative. It derives from anticipation of the problems that will or would arise during implementation.

In the Implementation phase we are faced with needing to carry out our threats, as the United States did in launching *Operation Desert Storm*, or failing to do so, as when the United States could not convince the Europeans they should follow through on Cyrus Vance's threat to "turn Bosnia and Serbia into a wasteland." Hard questions of ethics and self-interest arise at this point that were glossed over when making the threat. Many who previously supported our position now change their minds and join those who were always against it in arguing that we should not carry out our threat. As in the case of the inducement dilemma, such opposing subcharacters may use cost-benefit arguments. Threats having failed, it may seem that the realistic thing to do is to abandon our position.

The point is that these difficulties are foreseen, accurately or inaccurately, and possibly with exaggerations in either direction, by those who, at the moment of truth, are faced with assessing the credibility of our fallback position. The more difficulties they foresee for us, and the more reasonable they think we will be in succumbing to them, the less credible our threat, and the more likely it is that we will be faced with the hard choice of put up or shut up.

The Argentineans were amazed when the British fleet set out to recapture the Falklands, having confidently foreseen that Britain's threat to do so was futile.

THE POSITIONING DILEMMA

The positioning dilemma occurs when I prefer another's position to my own. In card-table terms, I like the other's position column more than mine.

This is likely to occur when, under pressure of dilemmas, I have abandoned a position I occupied with others and accepted one that I find less preferred, leaving others with whom I shared the first position to continue to occupy it.

A player might abandon a preferred position in favor of one it likes less because of arguments based on realism. At its old position, the player faced dilemmas, which, unlike the players that still occupy that position, it came to consider insurmountable; however, now it faces a dilemma in arguing with remaining proponents of its old position because it prefers what it is arguing against to what it is arguing for. The player stands accused of insincerity and dishonorable conduct in failing to stand and fight for what it once believed in.

The accusation is one of failure to live up to principles the player previously defended. A common example is that of two friends. In their youth they supported left-wing causes. In middle-age, one continues to support left-wing causes while the other argues that the positions they once shared are unrealistic.

Feelings of guilt at having failed to live up to once-defended principles accompany this dilemma. These guilt feelings arise out of the dilemma. They are structural. We are wrong to think of guilt feelings as being caused by the abandonment of higher feelings such as altruism in favor of lower ones such as selfishness. The same guilt is felt in response to a positioning dilemma in which lower principles are abandoned in favor of higher ones. To see this, consider two middle-aged friends who thieved and took drugs in their youth. One has reformed, primarily on realistic grounds (deterred by society's punishments for such behavior), but also on the grounds that the behavior is wrong. His friend will accuse him of deserting lower principles for higher ones, and if his main reason for changing position is realism, he will feel guilty in face of this accusation.

The positioning dilemma is essentially a guilt dilemma. It is overcome by a change of principles (or a reassignment of weights between different principles) that eliminates the dilemma by making the new position preferred to the old one. The emotion accompanying this change is the zeal and enthusiasm of a convert, rationalizing negative attitudes toward the old position and positive ones toward the new. Former Communists become zealous defenders of free-market capitalism. Former drug addicts campaign vehemently against drugs.

Often such converts are respected for their supposed inside knowledge of the position they are attacking. Actually, their knowledge may be biased, having been through strong emotional pressures.

Modeling Northern Ireland Negotiations

To see how the positioning dilemma may affect peace operations, we will model the negotiations in Northern Ireland preceding the talks that led to the Easter 1998 agreement. Here we will see a number of dilemmas at work, in addition to the positioning dilemma.

Table 5 shows the moment of truth between Britain, the Irish Republican Army (IRA), and the Unionists before the reconvening of peace talks in 1997. It focuses on the point when a new government under Prime Minister Tony Blair took over from the

	I	U,B	t	d
IRA/SINN FEIN	1	3	2	
Cease fire	[?]	[?]	☐	☐
Disarm	☐	[?]	☐	☐
BRITISH GOVERNMENT	2	1	3	
Admit Sinn Fein to talks	■	■	☐	☐
UNIONISTS	3	2	1	
Quit talks	[?]	[?]	☐	☐

LEGEND
■ means card is played
☐ means card is not played
I is position of IRA/Sinn Fein
U,B is position of Unionists and British
t is threatened future
d is the default future, which is the same as the threatened future
[?] means playing this card is not preferred
[?] means playing this card is preferred
# indicates preference ranking (1 is most preferred)

Table 5. Status of Northern Ireland peace talks when Tony Blair took over.

Conservatives, under which an IRA cease-fire broke down in early 1996. This cease-fire was meant to precede all-party peace talks. It broke down over the British government's insistence on backing the Unionist demand that Sinn Fein (the IRA's political wing) not be admitted to talks unless the IRA first handed in some of its weapons.

When Blair took power, the parties' positions were as shown in table 5. In column I, the IRA was offering to cease fire (but not disarm) if it could then be admitted to peace talks. In column U,B, the Unionist position, shared by the British, was that Sinn Fein be admitted to talks only if they disarmed. If this was rejected, the British and Unionists were threatening to go ahead with talks without the IRA (in fact, talks were ongoing, but marking time). Thus the threatened future consisted of the IRA neither ceasing fire nor disarming, the British not admitting them to talks, and the Unionists not quitting talks. The default future was the same.

In this table, as before, the numbers beside characters' names represent their preference rankings for the various futures, derived from the values they are trying to pursue in this confrontation. Number 1 is attached to the most preferred future, 2 to the next most preferred, and so on. From these preference rankings, some of the dilemmas can be read off. Others are indicated by question marks (i.e., a card has a question mark on it if there is some doubt if it would actually be played or not played).

For example, the distrust felt by each side toward the other is indicated by question marks beside the Unionists' "Not quit talks" choice in column I and the IRA's "Cease fire" choice in column U,B. This is because the IRA suspected that if talks took place on fair terms, as in column I, the Unionists would sooner or later quit such talks, while the Unionists suspected that the IRA would sooner or later break the cease fire even if, as in column U,B, they previously had given up some of their arms, which the Unionists in any case suspected they would not. (Even if they gave up most of them, the Unionists suspected, the IRA could always rearm.) This means that both the IRA and Unionists have trust dilemmas. Each also has a potential cooperation dilemma because each can envisage talks going in directions so unacceptable that they would prefer to return to violence or quit the talks.

These dilemmas gave the characters arguments to reject each others' positions. At the same time, each was inclined to regard its own position as second-best and to argue that, in light of the above dilemmas, the British should move with them to another position, one in which the other side was excluded from negotiations. For the Unionists, this preferred position was the default future *d*, the same as the threatened future *t*. For the IRA, it was the same as their position I, but with the Unionists provoked into playing their "Quit talks" card.

Britain's Dilemmas

What caused movement was the combination of dilemmas faced by the British. They had not only an inducement dilemma (they preferred I to *t*), but also a deterrence dilemma (the IRA, because they would find it impossible to retain internal discipline if they disarmed ahead of talks, preferred *t* to U,B).

It was the British deterrence dilemma that was decisive. Unless you can get rid of this dilemma, you are not in the game. Meanwhile, the IRA had neither an inducement nor a deterrence dilemma. It was their otherwise unsolvable deterrence dilemma that precipitated the Blair government's change of position. The Blair government accepted the IRA's position (i.e., that it should be admitted to talks without disarming). In table 5, the government shifted position from columns U,B to I.

The same deterrence dilemma faced the Unionists. It had the effect on them too of making them want to change position, but in a different direction. The Unionists used their deterrence dilemma, like their other dilemmas, to argue that they and the British should take the status quo (the default future d, the same as the threatened future t) as their position.

Why did the same imbalance not precipitate a change of position on the part of Blair's Conservative predecessors? The simple explanation, often given, is that they depended on Unionist politicians to keep their majority in Parliament, and hence stay in power. This might not have been enough to keep them aligned with the Unionists if a positioning dilemma did not lie ahead if they changed position. Consider the situation of these Conservative politicians. They preferred, for good reasons, the position U,B to the IRA position I. This meant they need not think about their majority in Parliament. Instead they could think of all kinds of principled reasons, such as not giving in to terrorists, for staying with the Unionist position. Blair's government, by contrast, went against its own principles in shifting position.

The Blair government went against its principles but embraced political realism. The solution U,B in table 5 was simply unrealistic. The Blair government might have tried to make it realistic by escalating other threats to the IRA, such as potential loss of U.S. and Irish support, followed by a heavier crackdown on terrorism. The Unionists suggested these extra threats, but the

Unionists wanted the threatened future, and all such threats were, in fact, things the Unionists wanted to happen for their own sake.

New Moment of Truth—Britain's Positioning Dilemma

When the Blair government shifted position, a new moment of truth emerged, as shown in table 6. Britain now shared the IRA's position. The IRA responded by calling a new cease fire, without disarming, giving a clear signal of support for what was now a joint British–IRA position. (Notice how this cease-fire, being reversible, acted merely as a signal of intent, not as an element of the implementation. Evidently it remained possible that the implementation, when it came, would not contain the cease-fire card.)

The threatened future was now that the talks would go ahead without the Unionists. The Unionists always said that if Sinn Fein was admitted to talks without having disarmed, they would quit. Blair now called the Unionists' bluff, declaring that the train, which under the Conservatives had threatened to leave without Sinn Fein, would now leave, if necessary, without the Unionists; however, in light of the Unionists' declared fallback position, the default future was now one in which the Unionists would refuse to join the train (i.e., it was the same as the new threatened future).

This new threatened future naturally changed the dilemmas parties faced. It was no longer the Unionists that preferred the threatened future to their own position; it was the IRA. The Unionists now faced an inducement dilemma, not wanting to accept their opponents' position I,B. Yet the Unionists preferred

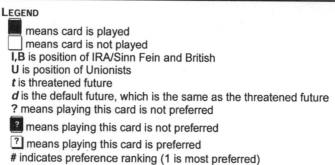

	I,B	U	t	d
IRA/SINN FEIN	2	3	1	
Cease fire	?	?	■	■
Disarm	□	?	□	□
BRITISH GOVERNMENT	2	1	3	
Admit Sinn Fein to talks	■	■	■	■
UNIONISTS	2	1	3	
Quit talks	?	?	■	■

LEGEND
■ means card is played
□ means card is not played
I,B is position of IRA/Sinn Fein and British
U is position of Unionists
t is threatened future
d is the default future, which is the same as the threatened future
? means playing this card is not preferred
[?] (black) means playing this card is not preferred
[?] (white) means playing this card is preferred
# indicates preference ranking (1 is most preferred)

Table 6. Status of Northern Ireland talks after Blair shifted position.

this position to *t*, because the latter would have left important decisions to be made in consultation between the British government and their sworn enemies. If the train was going to leave, they had to be on it.

In addition, they now faced a deterrence dilemma. Their position, U, which they now preferred to *t*, remained unrealistic. It was still the case (more so, in fact, than before) that *t* brought no pressure on the IRA to concede. The IRA preferred *t* to U even more than before. This deterrence dilemma gave the British an argument for refusing to accept U, even though in doing so the British had a positioning dilemma. (The government actually preferred U to its new position I,B.)

To overcome their now-pressing deterrence dilemma, Unionist militants argued in favor of escalation (i.e., campaigns of disobedience and disruption). Unionist leaders, however, saw little future in becoming lawbreakers and allowing the IRA to claim the mantle of respectability.

Other dilemmas as before faced the IRA and the Unionists, if their positions were to be accepted. The IRA and the British now had no deterrence dilemma; they were able to put pressure on the Unionists to give in. Nor did the IRA have an inducement dilemma.

The British did have an inducement dilemma, and worse, a positioning dilemma. In talking to the Unionists, the British had to argue against a position they preferred in favor of one they disliked. Their argument could not, therefore, be based on any supposed virtues of I,B as compared to U. They had to admit that U was better. Their point was merely that U was unrealistic.

The Unionists, meanwhile, could not rely on Britain's inducement dilemma. They might hope that the new threatened future *t* in table 6 would prove unbearable to the British, who would eventually be tempted to break off talks with the IRA and resume them with the Unionists. This hope was outweighed by the fear that the Blair government would face no such dilemma. The anguished Unionists finally accepted the position I,B and all-party talks began.

How to Get Rid of the Positioning Dilemma

In this example, Britain eliminated its positioning dilemma by getting the Unionists, whose position they preferred to their own, to accept their position. The government did so using the argument that the Unionist position was unrealistic.

Undoubtedly, they must also have argued that their position, which they shared with the IRA, was not as bad as they had formerly painted it. As long as the Unionists continued to reject realistic arguments, British negotiators were under pressure to eliminate their positioning dilemma, not just by appeals to realism, but also by adjustments to their beliefs and values.

The emotion that drives such elimination is embarrassment at being unable to give meritorious arguments for something that you are prepared to accept for reasons other than its merits. This produces negative emotion directed at the character you are arguing with, together with rationalizations building up the merits of the advocated position as against the one rejected. In extreme form, this emerges in the emotions of a convert, who generally feels more strongly in favor of its new position and against its old one than those who never took the old position.

Subcharacter Conflicts Over the Positioning Dilemma

The friction involved in overcoming the positioning dilemma is simply the difficulty you find in overturning the beliefs and values by which you once justified your old position against your new one. Scientists accepting a new theory, for example, must go methodically through each of the old arguments to see why points they previously rejected are now acceptable. Similarly, converts to a new morality have to critically review old life situations in which they made wrong choices.

This is what you must do to justify your new position. Meanwhile, for subcharacters who think we should have kept to the old position, our positioning dilemma is welcomed as an argument against the new position. It may similarly be welcomed by those who favor a third position, as they can use the dilemma to point out that we are being illogical.

When negotiations break down and characters enter the Conflict phase of figure 2, each character generally splits into subgroups, within which confrontations take place between those who want to proceed with the fallback position and those who want to restart negotiations or otherwise flunk the Conflict phase. In such a confrontation, those whose position is "Implement our fallback position" may be accused of inconsistency on the grounds that until now they have advocated another solution, the one they were urging at the Climax phase.

For example, following the breakdown of the Vance–Owen initiative in 1993, Secretary of State Warren Christopher urged the European allies of the United States to support lift and strike, (i.e., lift the arms embargo and supply the Muslims with arms while conducting air strikes against the Bosnian Serbs). Lift and strike was the threat Vance and Owen had used unsuccessfully to try to get the Serbs to agree to Vance–Owen. The European response was, "We still support Vance–Owen. Why don't you?"

Wanting to continue to press for the Vance–Owen position, the Europeans accused Vance of inconsistency in having abandoned it. Owen himself argued against the air strikes he previously had used as a threat, saying, "You will not solve the problem at 10,000 feet." The Europeans used what had now become the Americans' positioning dilemma, their advocacy of lift and strike even though they preferred Vance–Owen, as an argument against the U.S. position. The same argument was used by others who, judging that Vance–Owen could not be revived, advocated withdrawal or other positions.

This example illustrates another point. What was the threatened future during the Buildup and Climax phases becomes a position to be taken up in the Conflict phase. The future that you were using as a threat, all the time hoping you would not have to carry it out, becomes a position that you may have to argue for against opposition from other subcharacters.

SUMMARY OF CHAPTER 4

I face a *deterrence dilemma* in relation to a character opposed to my position, who prefers the threatened future to my position. The problem I face is that my fallback position does not pressure this character to accept my position. I can eliminate my dilemma by giving up my position. Alternatively, I can escalate to a higher level of retaliation by thinking up new cards to punish the opposing character. Anger and indignation, leading to demonization of the opposing character, accompanies this creation of new cards.

Conciliation is another way of eliminating the dilemma. To pursue conciliation, I retain the important characteristics of my position while sweetening it until the opposing character prefers it to anything it can obtain from the threatened future. A tough-cop, tender-cop routine is a way of combining escalation with conciliation.

Certain subcharacters within our character may prefer us to abandon our position and accept that of another player. They will tend to use a deterrence dilemma in relation to that other player as an argument for abandoning our position, on the grounds that it is unrealistic.

I have an inducement dilemma if another's position (different from my own) is as good for me as the threatened future. This puts me under potential pressure to accept that position, rather than go to the threatened future. I can overcome this dilemma by finding reasons to prefer the threatened future to their position, but in so doing I give them a deterrence dilemma. I escalate the conflict between us.

To avoid escalation, each side must accept its inducement dilemma. It must try to eliminate it by working out a common position with the other side. This means using passionately presented rational arguments in the common interest.

Against an extremist opponent, the best strategy may be, first, to eliminate our deterrence dilemma by mixing conciliation with a threat sufficiently strong that they cannot hope to escalate their way out of it; second, to eliminate our inducement dilemma by preferring to carry out this threat rather than accept any extremist position; third, to negotiate an acceptable common position that will not now be extremist. Under such circumstances, cost-benefit analysis will be an inappropriate tool for evaluating the threatened future. Concepts such as honor, commitment, and adherence to principle are appropriate.

I face a threat dilemma if I would be tempted to defect from the threatened future, if it were implemented. Unlike the inducement dilemma (with which it may, but need not, coincide) the deterrence dilemma affects characters at the Climax phase because they look forward to how it *would* affect them if they moved to the Conflict phase and considered whether to implement a future (the threatened future) from which they are tempted to defect.

I face a positioning dilemma when I prefer another's position to my own. This is likely to occur when I have abandoned a position I occupied with others and moved to one I find less preferred, but more realistic. My dilemma will cause me to feel guilty when arguing for the new position against the old, unless I can change my principles sufficiently to prefer the new one. Having made this change, I will, as a consequence, tend to uphold the principles supporting my new position even more strongly than those who have always held it.

An alternative way for me to get rid of my positioning dilemma is to get those who still hold my old position to move with me to my new one. This is how the British government eliminated the positioning dilemma it faced when arguing with the Unionists in favor of admitting Sinn Fein to talks although the IRA had not yet decommissioned arms.

A positioning dilemma may face subcharacters of a character that has to put up or shut up (i.e., decide whether to implement its fallback position after negotiations break down). Subcharacters who previously argued for internal implementation of a peaceful solution now may find themselves arguing for implementation of a punishing policy that was used in the higher-level drama as a threat to pressure others into accepting the peaceful solution. Other subcharacters may accuse them of abandoning their principles in switching from a peaceful policy of the internal game to a punishing one.

CHAPTER 5

WHEN ALL DILEMMAS ARE ELIMINATED

What happens when all six dilemmas are eliminated?

We have seen that while any dilemmas remain, characters are under rational and emotional pressure to get rid of them; therefore, the situation is not fully stable (i.e., there is discontentment and pressure to change). This must continue until no dilemmas are left.

Is the situation then fully stable, with fully contented characters? This question is important for the commander of a peace operation that has at least a partial goal to bring about stability. The answer is yes, with several qualifications.

THE FINAL STATE

It is true that the final state, where there are no dilemmas that give rise to pressures for change, is stable; all characters agree on a single position, and no character or group has any temptation to defect from this position. This theorem of the final state is proven in the mathematical appendix. It can be made intuitively clear as follows:

- If all parties agree on a single position, then the absence of any cooperation dilemma ensures that no character or group has any temptation to defect from it.

- If there is not a single position, then the only way for characters to eliminate their inducement dilemmas without creating deterrence dilemmas for each other is to converge to a single position.

- If there is a single position and no cooperation dilemmas, then no other dilemma can exist except the threat dilemma.

Note that the threat dilemma can still exist; however, this is unimportant because if all agree on a position that they can trust each other to carry out, then any lack of credibility in what they threaten to do, if they do not trust each other, ceases to matter.

The final state is, then, a completely satisfactory solution, satisfying both emotion and rationality. Correct?

That is correct; however, we have said there are qualifications to bear in mind.

Tragic and Happy Endings

The first qualification concerns what is meant by satisfactory. A drama may end tragically as well as happily. (It also may end tragically in some respects, happily in others.) Both kinds of ending are satisfactory to the audience of a drama (everything is settled); however, a tragic ending is unlikely to mean satisfactory completion of a commander's mission.

We are really using the word satisfactory in a scientific and aesthetic sense. Whether the ending is happy or tragic, there is stability in the form of contentment (i.e., no character has anything left to hope for or to fear). In either case, this comes about partly through modification of initial hopes and fears during the course of the drama. At a tragic ending, broadly speaking, hopes have been destroyed and fears realized. At a happy ending, hopes have been realized and fears banished. Clearly, a commander wants stability at a happy ending, although perhaps only after hopes have been appropriately modified, rather than at a tragic one.

For an example of the tragic type of stable ending, consider the following generalized model of how countries escalate their differences until they gladly go to war. Let each implicitly or explicitly threaten to fight unless its political and diplomatic position is accepted. Let each then eliminate its inducement dilemma by rationalizing a preference for war as compared to the other's position. Then, to eliminate any possibility of the other responding to this by further escalation, let it take its rationalization so far that war itself becomes its position, any compromise with the other being considered worse than war. War has then become a totally satisfactory resolution; that is, it is now a shared position from which no group of players is tempted to defect. Pursue this reasoning further and we find players escalating the type of war they consider better than any of the opponent's positions until they are committed to Clausewitzian total war with its absolutist demand for unconditional surrender.

The process in which friends and neighbors from different ethnic groups start to murder each other, as in Bosnia and Rwanda, may be explained as another kind of tragic ending. Each group knows, from tales handed down, that the group its neighbors belong to has in the past massacred them. Each group hears officially denied rumors that this group is arming to give itself the capability of doing it again; therefore, each group takes the precaution of arming itself. Finding now that its fears are confirmed, each group fears that the other now prefers to massacre it, and hence rationalizes a change in its attitudes so that it prefers to massacre the other first. The future, "Each tries to massacre the other before it can be massacred itself," is now necessarily interpreted by any unit within a group as meaning "We succeed in massacring them first." Unless and until this unit is massacred, this is the logical way for it to interpret this future. Next, the rationalizations each group has used to justify its preference for massacring the other, based on handed-down stereotypes, means that it now prefers this to any compromise or settlement. Hence, "Each tries to massacre the other" has moved from being almost the worst outcome for either to becoming a tragic "totally satisfactory resolution."

In general, escalation to a totally tragic ending happens in the following way. Each party has an inducement dilemma, which it solves by rationalizing a preference for the threatened future as compared to the other's position. Each party then has a deterrence dilemma, which it solves by demonizing the other and thinking up credible cards that make the threatened future worse for the other. Each then has an inducement dilemma, which it solves..., and so on. This cycle of rationalizations continues until "Each doing the worst it can to the other" becomes a shared position elevated by each above any possible compromise.

To break this cycle, each side needs to start solving its inducement dilemma not by preferring conflict to compromise, but by suggesting modifications of its own or the other's position to create a single position both can share. This is what happens when negotiations proper start (e.g., during the 3-week conference in Dayton, OH, that led to the Bosnia accords). Parties' thoughts are then directed toward agreeing on a single position, although they may not succeed. Escalation, on the other hand, tends to go through its successive stages when the parties are separated, lobbing pronouncements and symbolic actions at each other from a distance while addressing either their own constituents or interested third parties. It is when parties are not formally negotiating that there is need for escalation to be controlled and thought given to the construction of a single positive position. Analyzing how to do this may be the main contribution of confrontation analysis.

Dealing with the Details

A second qualification to bear in mind in interpreting "totally satisfactory resolution" is that the degree of resolution of a conflict is relative to the model we are using. Zooming in on a model to see more detail uncovers potential disagreements that disappear when we zoom out.

This is literally true of boundary disputes: the more we zoom in on an agreed boundary, the more potential disagreements are revealed about where exactly it should run. It is also true of agreements on matters such as cessation of violence, economic arrangements, prisoner exchanges, and so on.

With this in mind, reconsider figure 2, where characters cycle between the Buildup and Climax phases until either they fall into conflict or reach a single, totally trustworthy position. If the latter, they look at the details (in the Resolution phase) to see if their agreement is really stable (i.e., do they mean the same thing and can they trust each other). If the answer is yes, they proceed to the Implementation phase.

This is correct in theory; but in applications, is it realistic to expect a single, totally trustworthy position? Imperfect agreements reached with much distrust remaining seem more likely. Our answer is that imperfect, mistrustful elements of the agreement are hidden by zooming out. This is realistic. Parties, to reach agreement, deliberately use ambiguous or over-general formulations; therefore, we model them with card-tables where much detail is covered over by a few, general cards until all that is left is a single, totally trustworthy position. That reflects the methods of real-life negotiators.

Nevertheless, you must look into the detail you have covered up. That is, what parties do in the Resolution phase. They do it to check that the overall understanding they have reached is sufficiently sound and reliable, not to settle all details, which would be a never-ending task. Card-table models can be used to see the details of an agreement, as well as its overall shape. This is accomplished in the following steps:

- Start with a simplified model, representing parties' general positions, to deduce conclusions about dilemmas and dilemma-elimination.

- Subject your conclusions to exploration and criticism by adding cards and characters to the model, particularly any that seem likely to overthrow the conclusions. Assess their effect on how parties, at the moment of truth, may sum up issues in the complicated model in terms of another simple model, perhaps different from the first. A principle tool for such assessment will be analysis of subconfrontations between subcharacters. From your new simple model, deduce new conclusions about dilemmas and dilemma-elimination.

- Criticize your new conclusions again by adding details to the simple model. Continue until you have a satisfactory model.

Technically, the technique of card-table modeling can incorporate any number of cards and characters. The reason for using a simple model is not practicality; it is realism. Simple models are realistic because the characters themselves, at a moment of truth, must and will use a simple common reference frame to feel sure they understand each other. To model the frame in terms of which they reach agreement by, a complex model would be unrealistic. Complex models, as complex as possible, nevertheless should be built. They are built to model not the moment of truth, but the process by which characters look into the adequacy of their simple models.

Example of a Detailed Model (Northern Ireland, 1993)—Context Cards

To illustrate, table 7 shows a complex model used to represent and explore the position supposedly taken by the British and Irish governments in the Downing Street Declaration of 1993.

The model not only represents the cards that, under our interpretation of the two governments' position, should be played by the various parties in negotiations, it also shows a number of cards that (under the position) it was assumed would be played by external characters. These form part of the assumed environment of the negotiations, and are placed below the others in a sectioned-off part of the table headed "Context."

In this Context section, the symbol ~ is put against cards about which either assumption ("will be played" or "will not be played") could be made. Above the context line, the same symbol is put against cards where the Downing Street declaration took no position.

Note that filling in and interpreting a table like this requires detailed knowledge. This is useful; but it also underlines the limitations of a complex model. It is useful in assessing the impact of the details that underlie general statements and raising questions that might be important. We can take a comprehensive look at a mass of detail before summing it up in a simple model.

It underlines the model's limitations by showing why a complex card-table will be an unrealistic model of a common reference frame. Characters cannot assume that such details are common knowledge between

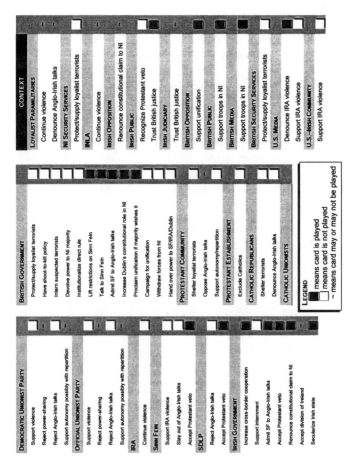

Table 7. Northern Ireland options in 1993-1994, showing in detail the position taken by the British and Irish governments in the Downing Street Declaration.

them, hence they cannot put them into a model of what they are sure each other knows. They know that each other's knowledge of details varies, with each appreciating the details of its own situation much better than the other's.

Implementing an Agreement—the Role of Subcharacters Within a Character

How then are details settled that have been glossed over by a simple, overall agreement? Each character consists, in principle, of an organization, and each of the chiefs that have accepted the agreement must order it to be implemented within their organization.

This is another reason why the overall agreement cannot include much detail. High-level officials cannot decide all the details of a complex operation; they must devolve them. Hence the orders given to implement an agreement, and so the agreement itself, must be simple and general.

Do lower-level decision-makers then simply fill in the details in accordance with their local knowledge? Not exactly. What happens is that the resolution of a conflict generally gives rise to a set of subsidiary conflicts, as it is implemented through confrontations between lower-level characters within each organization. Some of these confrontations are within organizations, some between them. When we model the overall agreement, we must realistically try to assess how such sequences of lower-level confrontations will work out in practice, just as the characters will try to assess the same thing at a moment of truth.

For example, in 1993 Serbian President Slobodan Milosevic, a player on the Serbian side who at that time favored acceptance of the Vance–Owen plan, recommended it to the Bosnian Serbs on the grounds that it could not actually be implemented. It would leave Serbian gains more or less untouched. However, this view was not generally accepted on the Serbian side, so it did not effectively become the Serbian view. The incident shows both how characters attempt to realistically assess what different agreements would mean in practice and how a character's views arise out of confrontations between subcharacters, here Milosevic and the Bosnian Serbs. (Silber and Little, 1996, p. 279).

Often an agreement looks forward to the conflicts that will arise in its implementation by setting up a subsidiary organization to help by handing out rewards and penalties, as was done following the 1995 Dayton accords.

There may be problems when a character in negotiations has insufficient power to order the implementation of what it has agreed. In modeling terms this is not a problem. It is simply a matter of correctly describing the cards given to characters. If a character cannot bring about a cease-fire, it should not have a "Cease fire" card. Its card should be called something else, perhaps "Recommend cease fire." Names should indicate characters' actual powers. This again reflects how characters themselves will model the situation at a moment of truth.

Unforeseen Contingencies

A third qualification to bear in mind when describing a resolution as totally stable is that it only reflects characters' projections of a possible future, to which the actual future is unlikely to conform.

Partly this is because of the previous point. Plans must be realized through sequences of confrontations between lower-level subcharacters, and the results of those confrontations, whether resolutions, conflicts, or defections, cannot be foreseen in detail. Their total effect may make implementation of an agreement better or worse than foreseen.

We do not know the future, so the plans we agree on are not generally fulfilled, even when we honestly try to fulfill them and all lower-level conflicts are resolved. Unforeseen contingencies arise, requiring the agreement to be interpreted to apply in circumstances it never envisaged, so that it may need renegotiating at various levels. This provides yet another reason for an overall agreement to be simple and general, as it must provide a framework within which such adaptations can be made. A clear yet general agreement allows flexible responses to changing circumstances while assuring subcharacters that the agreement remains in force, so that if they implement it they can continue to expect cooperation from each other and support from higher authorities. Agreements that go too far into detail run the risk of being seen too obviously not to apply when circumstances change.

For example, the history of the United States Constitution, or other constitutions, shows how a simple, general agreement is clung to, being reinterpreted when necessary, because the framework it provides for subcharacters' activities is too important and valuable to lose.

We have said before that an agreement that must survive under circumstances it did not foresee needs to be entered into in a positive spirit of goodwill, to reassure each subcharacter that others will strive to reinterpret and renegotiate it in ways that respect their interests.

WHEN CONFLICT RESOLUTION BREAKS DOWN

The conflict resolution process by nature may always break down. It can fall into the Conflict phase of figure 2, the phase where parties prepare to implement their fallback positions, having failed to bring about sufficient change to enable them to move from the Climax phase back to the Buildup phase.

In peace support operations, conflict normally means resort to or continuation of armed violence. Such violence may not involve our own force. Others may do the fighting, such as when UN peacekeepers withdraw from a deteriorating situation (cf. UNEF's 1967 withdrawal from the Middle East). Armed violence normally results. In other kinds of confrontation, such as between different components or coalition partners on our own side, conflict normally means a standoff with failure to cooperate, causing a diminution in operational effectiveness (e.g., failure to share intelligence or meet minimum interoperability requirements leading to loss of life through friendly-fire incidents, equipment malfunction, or inability to give timely support).

In disaster relief or humanitarian operations, breakdown of conflict resolution may mean similar standoffs, reducing operational effectiveness through failure to cooperate.

The Details of Implementing a Threatened Future

When parties in the Conflict phase, having failed to make their threats sufficiently credible or awesome to deter others, are faced with having to carry them out, the emotions aroused may be strong enough to convert the threatened future itself into a formal resolution of the conflict, as in the case of the tragic ending discussed above. Such a tragic ending is a complete, dilemma-less, totally satisfactory resolution; however, we do not say formally that it goes through the Resolution phase of figure 2. We say it goes through the Conflict phase.

We say this to maintain an essential difference between the Resolution and Conflict phases of figure 2. Characters in the Resolution phase sit down together to examine the details of their shared position. In the Conflict phase, each character meets separately to discuss within itself (i.e., each holding discussions between its own subcharacters) the implementation of its fallback position. The characters at the level of the confrontation as a whole do not communicate. (Their attempts to communicate would be defined as attempts to go back to the Climax phase and restart negotiations.)

Can the characters in a three-or-more person confrontation get together and communicate in subgroups?

This too can be outlawed as a matter of definition. Characters must have a common view, as part of their common reference frame, regarding which subgroups of characters would get together if the threatened future goes into the Conflict phase. That common view should determine the definitions of characters and subcharacters (i.e., we should define communicating groups of characters as subcharacters of a single character).

All this is a matter of formal definitions and modeling procedures, but it reflects important real-world distinctions.

For example, after the breakdown of the Vance–Owen initiative, the United States and European countries conferred as to the next step they should take, finally deciding not to implement their fallback position (lift and strike); likewise, the Bosnian Serbs conferred with the Serbian government. The subconfrontation between them entered its own Conflict phase as Milosevic angrily cut off military supplies to the Bosnian Serbs, although afterwards he resumed supplies. This means that in analyzing the climax of the 1993 confrontation over Vance–Owen, we should regard the Serbs as one character, comprising the Bosnian Serbs and Serbian government as subcharacters, and the West as another character comprising the U.S. and European governments. This reflects the way the parties perceived which parties they belonged to (Silber and Little, 1996, p. 287).

The Conflict Phase Is Game-Theoretic

From this we get a useful theoretical insight. The decision problem facing characters in the Conflict phase, after they have absorbed the preference and belief-changing impact of emotions and rationalizations generated at the Climax phase, and provided they do not decide to go back to the Climax phase, is essentially game-theoretic. All the methods of present-day game theory (which has, in the main, turned its back on the tendency to explore and pay attention to dilemmas) are applicable in principle at this stage.

There are some qualifications to this. The internal confrontations between subcharacters are not at all game-theoretic. Moreover, characters may look forward to communications in later confrontations with parties they are now cut off from by the breakdown of negotiations, and their preferences in their current situation may be influenced by the positions they foresee in such later confrontations.

For now they must make a decision in the manner assumed by game theorists (i.e., without communication, and hence by choosing the best course of action within a fixed, given framework of beliefs and preferences). They must do this whether or not the threatened future has become a tragic but nevertheless final resolution caused by emotions generated in the climax (i.e., whether or not they still face dilemmas in their relations with other characters). If they do, the situation requires that they ignore them and simply pursue the dictates of instrumental rationality.

The traditional Clausewitzian military mission is best understood in this sense. Rapoport (1968, pp. 69-77) points this out, while nevertheless emphasizing the limitations of traditional quantitative game theory as a decision-support tool. Clausewitz assumes that military specialists are given a mission to carry out against enemy forces if political negotiations have broken down and changed national attitudes have brought about passionate hatred between nations. The mission's objectives are political in that its aim is to use force to create a situation where resumed political negotiations will be more favorable to the nation. Its conduct is non-political. During this military interlude, negotiations with the enemy concerning the issues being fought over are out of the question, although tactical-level negotiations may be carried on over such issues as cease-fires to remove bodies, local surrenders, and declarations of open cities. Meanwhile, relations between the different players on our side, including the government, various components of the military, and our allies, are not at all game-theoretic. Confrontations between their differing views are supposed to lead to willing cooperation in meeting the war's objectives.

The Fog of War

Because the Conflict phase is game-theoretic, players in this phase encounter the fog of war in Clausewitz's sense. This is because Clausewitz's fog of war has two causes not present in negotiations when we are trying to guess at other parties' true attitudes and beliefs to pressure them into accepting our position.

The first cause of the fog of war, absent during negotiations, is that physical actions and operations are being undertaken for their physical effects; therefore, they are undertaken secretly. If the reason for undertaking a physical action (e.g., a bombardment) is to send a message, you will not try to make it secret. If you do, the message may not be received. When the aim of the action lies in its destructive physical effects, keeping it secret has the advantage of preempting countermeasures. Enemy secrecy causes fog.

In addition, physical operations tend to meet unforeseen contingencies, causing them to unfold in ways we did not expect. This creates more fog as units' ideas of what is happening to each other diverge.

The second main cause of fog in the Conflict phase is that, because we are not negotiating, we have no messages to interpret. Someone sending you a message is trying to make you understand something, and revealing themselves. There is a problem of what to believe or not believe. That problem may be called the fog of confrontation. It is different in kind from the fog of war, where the problem is one of having no messages to interpret.

FLUNKING AND DEFECTING

We have discussed the Resolution phase and the Conflict phase of figure 2 as if they are followed by implementation, respectively, of the current common position and threatened future.

They may be. But we have said these are not the only two possibilities. A third is interruption, caused by new exogenous information overturning the characters' common reference frame (e.g., Pearl Harbor upsetting the frame of Anglo–United States negotiations). Apart from this, two other possibilities are the betrayed resolution and the flunked conflict. The first consists of one or more characters defecting from the common position (e.g., the Nazis' betrayal of the Nazi–Soviet pact). The second consists of one or more defecting from the threatened future (e.g., the West's failure to punish Serbian rejection of the Vance–Owen plan).

There are, therefore, five ways a confrontation may end, usually setting the scene for new ones. Of these, the decision to flunk is, we have said, game-theoretic (i.e., instrumentally rational). The preceding Climax phase may have radically altered our attitude. Also, we may look forward to future confrontations in which our credibility may be affected by the actions we take now. However, given these considerations, we flunk because we consider flunking to be the best thing for us in light of our predictions of what others will do.

Consider the classic case of a terrorist who has threatened to blow up the plane he is on if his demands are not met. They are not. It is put-up-or-shut-up time. He now must ask himself, given the emotional confrontation he has been through, and thinking of the future he must look forward to if he gives in, whether he really prefers death. Given his operative values and beliefs at the time, the decision he makes will be instrumentally rational.

Is the decision to defect from an agreed position also game-theoretic? In principle, yes. Having gone through the Resolution phase where the parties confirm their agreement with each other, each must then separately decide whether to keep its part of the agreement.

Logically, this must be so. There must be some point at which parties decide whether to keep to their agreement, otherwise there would never be any doubt about this, as there clearly is. Yet the game-theoretic nature of this decision is not so clear as in the parallel case of the Conflict phase. For this there are several reasons.

In the Resolution phase, parties generally have used positive emotion and common-interest argumentation to make themselves prefer to keep to the agreement rather than pursue any temptation to defect. In this they may have been successful, particularly if the discussions have involved subcharacters within each main character. The result is that when they make their separate, game-theoretic decisions as to whether to defect, they often find they now prefer not to.

Secondly, during the Resolution phase parties may have agreed to an implementation plan instituting incentives for lower-level decision makers to stick to the agreement; they will have done so in order to eliminate temptations to defect. (For example, the Dayton accords invited into the area the NATO Implementation Force.) By affecting the incentives of subcharacters within each main character, this again makes the main characters prefer, on the whole, to stick to the agreement.

Thirdly, parties who have reached an agreement often look forward to a relationship where they will benefit from cooperating with each other in future confrontations; therefore, it is important for them to build up and maintain positive credibility (i.e., the belief that they will not defect from agreements). This gives them a reason not to defect from the present agreement.

To see the importance of the last two factors, compare two different scenarios in which a British colonel gets local militia forces to agree to remove its weapons from a certain area. In Scenario 1 the British commander will, immediately after the agreement, withdraw his battalion to be replaced by forces of another nationality (e.g., Russian), with which British forces have poor liaison, and which the local militia think are likely to be more sympathetic to their side. In Scenario 2, the British battalion stays to oversee the keeping of the agreement. We may suppose that under Scenario 1, the British colonel and local militia commander, knowing what will happen next, go through a Resolution phase in which they expend much effort and goodwill convincing each other their agreement will be kept; however, no matter what they say, it is more likely to be kept under Scenario 2.

SUMMARY OF CHAPTER 5

When all dilemmas have been eliminated, it is necessarily the case that all characters agree on a single position and can be trusted to implement it (see Appendix). In this sense, a totally satisfactory resolution has been reached, subject to a number of provisos. First, "satisfactory" is meant in a scientific or aesthetic sense, even though in this sense a tragic ending with hopes destroyed is as satisfactory as a happy one in which hopes are fulfilled. In real life, we generally prefer happy endings. These are brought about by managing the resolution process so that characters attempt to overcome inducement dilemmas by negotiating a single position, rather than by escalation. Secondly, resolution is model-dependent. A more detailed model, such as is examined by the characters in the Resolution phase, may reveal disagreements that require renegotiation of what has been agreed. The card-table modeling technique may be used to examine the details of an agreement. Thirdly, most details are settled, not in the agreement itself, but in the course of confrontations between subcharacters during implementation, confrontations that may or may not be resolved satisfactorily. Fourthly, we cannot know the future, so implementation may turn out unsatisfactorily because of contingencies not foreseen in the agreement. Finally, a perfectly satisfactory resolution may be reached in terms of the common reference frame and characters' communications, with one or more characters nevertheless intending or later deciding to defect from it, after having deceived the others into thinking they were trustworthy.

The same considerations apply, in a different way, if the confrontation ends not in agreement, but in conflict (which for defense forces usually means armed conflict). In the Conflict phase, characters separately look into details and confer between their subcharacters as to how to implement their fallback positions. Detailed consideration may cause them to change their mind, and either return to the Climax phase or simply not carry out their threats. When they do attempt to implement the threatened future, it usually turns out unexpectedly, leading them into fresh confrontations.

CHAPTER 6

THE FRONT-LINE PLAY: A DRAMATIZATION OF A CONFRONTATION ANALYSIS

M ost of this chapter consists of a play. It was written by the author under a contract with the United Kingdom Defence Evaluation and Research Agency (DERA). It was performed at a seminar on confrontation analysis held at DERA (Portsdown West) on February 5, 1998. Its object was to convey as vividly as possible how a commander tasked with a peace mission, and knowing nothing about confrontation analysis, might use the method to formulate and begin to implement a strategy for winning an Operation Other Than War (OOTW). It is reprinted here with the permission of DERA to give the reader an example of the formulation and implementation of a confrontation strategy.

A fictionalized peace mission is used, although with obvious resemblance to real ones, to focus attention on the process of analysis and strategy formulation rather than on the details of a particular case. The characters are British because the underlying research was done with British defense forces.

WOULD IT BE DONE THIS WAY?

In the play General Deloitte, commanding a combined joint task force, analyzes his situation by calling in a confrontation analyst on his staff. The general knows nothing about the method; it is something he has just heard about. He analyses the immediate problem at his level, then asks the analyst to further analyse the following items:

- Grand strategic problem that gave rise to his being tasked with this mission

- Other problems at his own level, both simultaneous with the immediate one and to be expected as the campaign develops (i.e., the linked sequence of confrontations)

- Internal problems of coordination between different forces in his coalition and different components

- Problems at lower levels, consisting of tactical confrontations that should be resolved in ways that support and are supported by his operational strategy.

What can be said about the organizational arrangements under which the general does this? Although better than doing no confrontation analysis, they are imperfect. The general and his chief of staff should have learned the method in normal training procedures. They would not then depend on the analyst for guidance at every step, as they do in the play.

As he does in the play, the analyst would depend on them for the strategic decisions that would underpin his more detailed, lower-level models. For these models he would need more staff than himself, and the models would need to be developed through interactions with responsible commanders at all levels.

In the next chapter we explore the principles of constructing a confrontation strategy. While reading the play, remember that here we are showing the fundamentals of the method by taking the case of a commander who has encountered it for the first time and is learning it at the same time he is using it. This is not meant to be ideal.

THE PLAY

[*The scene is a partitioned area of the banqueting hall of a tourist hotel near Morubwe, the capital of the North African state of Morya. The hotel has been requisitioned for HQ UNFORMOR (United Nations Force in Morya). The particular area we are in is the office of the commander (UNFORMOR), Major-General Eric Deloitte, CBE. It contains the general's desk, a large hotel dining table with chairs set round it, and a flipchart.*]

[*Commander UNFORMOR is British because Britain has been designated the framework nation for the UN's intervention in Morya. Under him is a British brigadier commanding the Anglo-Egyptian land forces, a U.S. Air Force brigadier-general commanding the air forces, and a colonel of the Armée Francaise commanding the aviation force, a battalion of helicopters available to support the land forces.*]

[*General Deloitte, a determined-looking, middle-aged man in combat dress, is standing looking discontently out of the window. Seated at the table is his chief of staff, Brigadier Ray Jones. The chairs are drawn back from the table and papers are scattered on it. Evidently a meeting has just finished. On the flipchart are scrawled the names of some of the main protagonists in the Moryan drama]:*

MORYAN GOVERNMENT

ISLAMIC REVOLUTIONARY ARMY (ISRA)

GOVERNMENT OF PELUGYA (long common border, supplies ISRA)

FRANCE (tends to support Moryan govt)

EGYPT (strongly against ISRA)

ARAB COUNTRIES (some support for ISRA)

[Brigadier Jones is busy at the table sorting through papers and making notes.]

DELOITTE: Is everything ready for the press conference?

JONES [*still writing notes*]:   Yes sir. At the airport. Fourteen hundred. We should leave 1330.

DELOITTE [*irritably*]:   What's the right way to handle this? It's no good, you know, Ray. We're not there.

JONES [*raising his eyebrows and turning to look at the General's back.*]:   Really, sir? This morning's meeting was the most thorough review of the situation we've had. I'm just writing up the notes for you. Each component reviewed the situation from their viewpoint. Intelligence gave a good overall assessment. We know with fair accuracy what's going on, how the relief convoys are getting through, etc.... As to the press,

I've seen you handle them. I'm sure you can refrain from giving away anything that'll further inflame the situation.

DELOITTE [*disgustedly*]: That, precisely, is not the point. This press conference should be my first move in a planned campaign to reach my objectives. It shouldn't be a matter of giving nothing away, of giving no hostages to fortune. But I've nothing resembling a plan. [*Turning to face Jones*] That's the trouble, Ray. Soldiers know how to fight battles. I can plan for a battle. But this, this is not anything resembling a battle. The objective here is to get your way without the use of force. We are not trained to do that. That's a politician's job.

JONES [*wisely*]: Peacekeeping isn't a soldier's job, but only a soldier can do it.

DELOITTE: Who said that? Never mind, it's quite right... [*Goes back to gazing morosely out of the window. Then starts again, waving his hand at the flipchart.*] You know, Ray, I've a hunch the Moryan government, not ISRA, will be the one that refuses to cooperate in peace talks. Apparently we still have not got the President to agree to a time and place for an emergency meeting with ISRA. That's why they called me out of the meeting half an hour ago. I told them to keep trying.

JONES [*surprised*]: The government make difficulties? Why should they? Our intelligence is that they're under the most pressure to end the fighting. They have most to lose from it continuing.

DELOITTE: That *was* the position. My hunch is that things have changed. I suspect they think all this world publicity against ISRA has tipped the balance in their favor. They think it'll force us to intervene on their side. So, the worse things get for them the better. That, I suspect, is their thinking.

[*Jones shakes his head disgustedly.*]

DELOITTE: Meanwhile, ISRA will continue their campaign, regardless what the world thinks. And I'm supposed to do something about it. Without using force. But what? [*Musingly*] You know, I'm trying to remember something I heard recently. About a new thing called confrontation analysis. That's what I need. After all, that's what PJHQ has thrown me into. A confrontation. That's the term for it.

JONES [*hesitantly*]: Sir...

DELOITTE: Yes?

JONES: We've someone on the HQ staff who's trained in confrontation analysis. He is a civilian analyst in OA called Bright, Mark Bright. DERA sent him along. We haven't used him in that capacity. OA have had their hands full programming databases for us. But I could get him in to tell you about it if you like.

[*The Commander looks at Jones thoughtfully, as if his mind were on other things. Then he shakes his head as if to reproach himself for indecision, and glances at his watch.*]

DELOITTE: Send him in.

JONES: Yes sir.

[*Jones gets smartly to his feet and leaves the room. The general returns to his morose vigil at the window.*

Almost at once Brigadier Jones returns leading Mark Bright, an intense and slightly supercilious young man in civvies. The general turns round. Bright, without actually saluting, stiffens and inclines his head as if to indicate military readiness.]

JONES: This is Dr. Bright, sir.

DELOITTE [*regarding him curiously and somewhat humorously*]: Sit down, doctor. [*Bright does so.*] Tell me about this new discipline DERA has trained you in.

BRIGHT [*promptly*]: Yes sir. Confrontation analysis is an OA tool. You analyze situations where each party takes a position, meaning a suggested solution, what it suggests everyone should do, itself included. Generally, parties' positions don't coincide. They're proposing different solutions. So there's a conflict of wills. As well as their positive positions, each party in a confrontation has a fallback position, what it says it'll do if its position is not accepted.

DELOITTE: That defines a confrontation?

BRIGHT: Yes sir. In a way confrontations correspond to battles. To win a war, you win a linked sequence of battles. To win a peace support operation, you could say, you fight and win a linked sequence of confrontations.

DELOITTE [*in a lively tone*]: Well, suppose I'm in a confrontation. How do I analyze it?

BRIGHT: Well, once you've decided on everyone's positions, you look for the so-called dilemmas facing each party. Essentially, dilemmas are credibility problems. For example, I have a threat dilemma if you think I would not carry out my threat. In other words, if you think I'm bluffing. I have a trust dilemma if I couldn't trust you to carry out my position, even if you agreed to it. There are four other dilemmas.

DELOITTE: They sound realistic enough.

BRIGHT: They're extremely realistic, in my experience, sir.

DELOITTE: Can you always find these dilemmas? In any confrontation you look at?

BRIGHT: They're guaranteed to exist, sir, unless and until all parties have agreed on a single solution they can trust each other to carry out. In other words, if there's no dilemma, there's no problem. If there's a problem, the parties are necessarily facing dilemmas. So by eliminating all dilemmas, you solve the problem.

DELOITTE [*bridling*]: You do, do you? On whose terms? Whose solution do you end up supporting?

BRIGHT: We have to make sure it's our solution that wins, don't we, sir?

DELOITTE [*loudly*]: And how the hell do we do that? [*He leans forward with his hands wide apart on the table looking down at Bright, who sits across from him.*]

BRIGHT [*quietly, looking down at the table*]: It's a matter of making a plan and following it. Winning a confrontation is like winning a battle, only different. The details are different, but the principle's the same. You exploit the other side's weaknesses and our own strengths. First off, you analyze the dilemmas everyone faces. Dilemmas are points where we can exert pressure on them, or, if we're not careful, them on us. Next, you make a plan to eliminate dilemmas in a sequence of operations that will bring the others into full compliance with your objectives. [*He looks up to find the general staring fixedly at him. He pauses uncertainly, then continues more quietly than before.*] It's a logical process, sir. Supposedly, it's as logical, or more so, than winning a battle. Of course, you've still got the fog of war to contend with. I mean, you have to make assumptions because you can't be sure of the facts, and your analysis is dependent on the facts you put in. But you've got to act. So you make the best assumptions you can, whilst being ready to change your plan, without vacillation or confusion, if your assumptions prove to be wrong... [*His voice falters as the general continues to glower at him.*]

DELOITTE: You're quoting Clausewitz.

BRIGHT: Well...

DELOITTE: That's straight Clausewitz.

BRIGHT: Yes sir.

DELOITTE: What does winning mean when you're in a confrontation?

BRIGHT [*without hesitation*]: Bringing others into full, willing compliance with your objectives.

DELOITTE [*turns his back and walks to look out of the window. After a pause*]: Okay, suppose I want to analyze the confrontation I'm in now. How do I do it? What do I need? [*Looking at his watch.*] How much can I do in one hour?

[*Bright looks appealingly at Jones, who returns his look impassively. The general continues to gaze out of the window, his back to them.*]

BRIGHT: Er, well, sir, I could take you through a broad, high-level analysis. I'd sort of ask you a series of questions and show what follows from the answers. You'd be responsible for all assumptions, and obviously decisions taken on the basis of the analysis would be yours. I'd be responsible for the process of analysis, showing you how to do it...step by step...

JONES [*imperturbably*]: Equipment needed?

BRIGHT: Just a flipchart. In an hour, we might...map out a broad policy, though probably it'd raise a lot of questions for intelligence to answer. We'd have to do more detailed analysis and checking later on. Then we could make detailed plans to be implemented at various levels and by various components. [*Looking anxiously at Jones while still talking to the general, whose back is still turned to them*] I'd have to start, sir, by asking what decisions have to be made or actions taken in an hour's time.

[*A long pause.*]

DELOITTE [*turning round briskly and coming to sit at the table*]: Of course. [*Waving his hand at the flipchart*]: There's your equipment. Ray, perhaps you'd better sit in on this. I may need your views on some of the...er...assumptions. [*To Bright*]: Let's go.

BRIGHT [*standing up awkwardly and going to the flipchart*]: Yes sir. Er...

DELOITTE [*sitting easily, his hands wide apart on the table*]: To answer your first question. I have to give a press conference at 1400 hours. I could just regard it as a nuisance, and give away as little as possible. That would be our usual style, I suppose. [*He looks mischievously at Brigadier Jones, who does not respond.*] Instead, I want to handle it as the first step in a PLAN…to achieve my OBJECTIVES. So, at 1400 I'll be facing reporters and cameramen from all the world's media. These are the people who've whipped up public hysteria in Britain, America, and Europe over the Friday massacre attributed to ISRA. Dead babies on television. All that kind of thing. Horrified the whole world. The Cabinet met to discuss it yesterday. The Defence Secretary contacted the CJO, who called me in. I got back from PJHQ early this morning.

BRIGHT: And you came back with a new mission, sir?

DELOITTE [*sardonically*]: My mission, if you can call it that, is to get the politicians out of trouble. Somehow or other, I have to stop ISRA massacring people, or at least, stop it getting into the news. Once again, we've been landed with what ought to be the politicians' job. But CJO overrode my objections. Said we've got to do something.

BRIGHT [*hesitantly*]: So, sir, your objective is…? How would you state your strategic objective?

DELOITTE [*after pausing to think*]: To get a cease-fire between ISRA and the government, with an agreement between them to start discussing a settlement. That's my objective. To be achieved without using force.

BRIGHT: If possible…

[*The general does not answer.*]

BRIGHT: I mean, it's possible the only way to get your way without using force is to be ready to use force, if necessary. Wouldn't you agree?

[*The general gives a short, bitter laugh. Brigadier Jones smiles wanly.*]

DELOITTE: You're right, in general. But you're straying into a conceptual minefield. Doctrine, so far as we've got one, distinguishes peacekeeping, where you don't use or threaten force, from peace enforcement, where you do. We're supposed to be here in a peacekeeping role. On the other hand, a peacekeeping situation is one where the parties are prepared to be peaceful. In Morya, they're not. So we're in a peace enforcement situation with a peacekeeping mandate. What do we do about that?

BRIGHT [*rising to the challenge, picking up a pen and folding up the flipchart to disclose a blank sheet of paper*]: I can't, of course, tell you what to do, sir. [*Jones raises his eyebrows at this naïve remark.*] All I can do is help you analyze the situation to see what's possible, that is, how far it's possible to get peace without threatening force. And if so, how. Now…

First Narrated Interlude

[*As he raises his pen, the narrator steps forward and lifts his hand. The three characters freeze in position.*]

NARRATOR: General Deloitte, assisted on request by Brigadier Jones, now answers Mark Bright's questions as to the positions currently being taken by the parties. This enables Mark to draw up this card-table. [*He folds up the blank sheet on the flipchart to disclose a sheet, shown here as* table 8.]

NARRATOR: As Mark explains to the general, he has modeled the confrontation by giving each protagonist certain cards to play or not play. The various parties' positions are then shown by listing, in separate columns, the cards each party proposes should be played.

So the UNFORMOR position, column U, is that both sides should cease fire and join peace talks. UNFORMOR then will not call air strikes against ISRA (although this is what the Moryan government wants) nor recommend UN withdrawal (which is what ISRA wants). Note the name given to the card, "Recommend withdrawal." The general can't actually decide on a

	U	I	M	*t*	*d*
UNFORMOR					
Call air strikes on ISRA	☐	☐	■	☐	☐
Recommend UN withdrawal	☐	■	☐	☐	☐
ISRA					
Cease fire	■	☐	☐	☐	☐
Join peace talks	■	☐	☐	☐	☐
MORYAN GOVERNMENT					
Cease fire	■	☐	☐	☐	☐
Join peace talks	■	☐	☐	☐	☐

LEGEND
■ means card is played
☐ means card is not played
U is position of UNFORMOR commander
I is position of ISRA
M is position of Moryan government
t is threatened future
d is the default future, which is the same as the threatened future

Table 8. Moryan government no longer wants a cease-fire.

withdrawal, he can only recommend it to his superiors. Hence the name.

ISRA's position, column I, is that the UN should withdraw, and therefore that the general should recommend this to his superiors. He shouldn't call air strikes against ISRA positions. ISRA and the Moryan government should be left to fight it out. ISRA expects to win the resulting civil war and install an Islamic regime.

The Moryan government's position, column M, apparently has just changed. It seems unwilling to attend the meeting the general has called, which indicates that it's now taking the position that there should be no cease-fire or peace talks. Instead, UNFORMOR should intervene on its side by calling air strikes against ISRA. Until now, while continually asking for air strikes, the government hasn't objected to the idea of a cease-fire followed by talks.

These are the parties' positive positions. But a confrontation consists not just of positive positions. It includes parties' fallback positions as well, the unilateral actions they say they'll take if others don't convincingly agree to their positions. If all parties carry out their fallback positions, we get something called the threatened future. This is shown in column t. It so happens that in this case it's the same as the default future (column d), the future to be expected if everyone continues their present policies without change. This is a future under which ISRA and the Moryan government will continue to fight while UNFORMOR limits itself to humanitarian aid. So UNFORMOR does not call air strikes nor recommend withdrawal, while

ISRA and the Moryan government neither cease fire nor join talks.

This is the future that has led, through one particular incident, to the present crisis. The general situation is that government forces are maintaining control of the center of Morubwe and other big towns, while ISRA control most of the country. ISRA are being supplied militarily by the neighboring government of Pelugya, which has a radical, fundamentalist Islamic government, while bombarding government-controlled areas, causing deaths which tend to get reported by the Western media. The particular incident which has now caused an international outcry and led to the general being told to do something is the bombardment of a crowded market, which caused sixty deaths.

And the background to all this?

Three years ago ISRA supposedly won an election. But the secular government of Morya, headed by President Saldin, hoping for Western support, annulled the election results. ISRA thereupon began a bloody revolt and Saldin asked for Western help. The UN approved the sending of UNFORMOR with a mandate not to take sides, but to protect humanitarian relief missions while encouraging the parties to agree on fresh, UN-supervised elections. ISRA have rejected this plan, arguing that they have already won elections. President Saldin has gone along with the plan until now in order keep Western support, but he really wants the West to intervene militarily and help him defeat ISRA. Meanwhile, the UN has banned military flights and imposed sanctions against Pelugya for supplying ISRA, while UNFORMOR has brought in aid to the populations of the cities and sent convoys to aid rural areas, entering into negotiations with ISRA to allow them passage. Thus it has built up cooperative relations with both sides in pursuit of its humanitarian objectives.

[*The narrator waves his hand, causing the characters to come back to life. He walks off the stage and leaves them to it. The general stands up and leans forward across the table, gazing intently at the flipchart.*]

Resumption of the Play

BRIGHT: So it seems this [*indicating the flipchart*] is the moment of truth in the drama between you, ISRA, and the Moryan government. Obviously, there are other dramas. There's the higher-level, Grand Strategic drama between governments—the U.S., Britain, France, the Moryan government, Pelugya, various Arab governments. That's the drama that led to the formation of UNFORMOR, that is, it's the drama that defines your superior's intent at two levels above.

DELOITTE [*sardonically*]: If there is such a thing.

BRIGHT: I mean, sir, that the requirement for a commander to understand his superior's intent, two levels up, is not very clear in this kind of UN peace operation. In this kind of case superior's intent is the resultant of political interactions between governments. So understanding that grand strategic drama is essential. So is understanding the lower-level, tactical dramas that are also going on between the forces you command and other forces, as well as NGO's, aid organizations, communities, and so on. You have to direct those lower-level dramas to implement a whole, cohesive plan for achieving your objectives.

JONES [*glancing at his watch*]: We don't have time for all this, Dr. Bright. The general meets the press at 1400.

BRIGHT: I understand that, sir. Analyzing those other dramas will have to be done later, in follow-up work. [*Gesturing at the flipchart*] This is the one we'll concentrate on now, but we have to remember those others in the background because we'll be making assumptions about them.

DELOITTE [*as if talking to himself*]: How do I get them to agree on column U?

BRIGHT: Ah. Yes, sir. First we have to analyze the dilemmas everyone faces.

DELOITTE [*with a quick gesture*]: Show me those.

BRIGHT: I'll have to ask you some questions about players' preferences. First…Isn't it true that both ISRA and the Moryan government prefer this, the threatened future (the same as the present, default future) to our position, column U? Isn't that correct? [*As Bright refers to each future, he points to the corresponding column on the flipchart.*]

DELOITTE: Hmm…I don't know about that. ISRA wants us out…Saldin wants us to intervene against ISRA…

BRIGHT: Right. The present future is not the position of either of them. But in the case of the Moryan government, surely it hopes, by milking the present situation, to get the UN to tell us to intervene against ISRA. Notice, here we're talking about its assumptions concerning the grand strategic drama.

DELOITTE: In that sense I believe you're right. Correct, Ray?

[*Brigadier Jones nods.*]

BRIGHT: So, taking into account those assumptions about the grand strategic drama, the government does prefer to continue with the current future, the same as the threatened future, rather than move with ISRA to our position. [*They both nod.*] Okay. Now, ISRA...Surely ISRA believe the current conflict will end in their victory.

JONES: That's why they want us out, Dr. Bright.

BRIGHT: Yes, in other words they'd most of all prefer this, their position [*indicating* column I]. But between continuation of the present conflict with us here but remaining neutral, and a cease-fire with peace talks (which is our position), they'd prefer the present conflict.

DELOITTE: Correct.

BRIGHT [*triumphantly*]: Then we have it. We face a deterrence dilemma. Fatal. We must solve it, or our position is untenable. Unrealistic.

DELOITTE [*sardonically*]: Presumably you're trying to say that because they prefer being where they are, they're under no pressure to accept our position.

BRIGHT: Exactly. Or rather, they prefer what we're threatening them with, which in this case happens to be the same as the present, default future, to our position. So, as you say, we're placing no pressure on them to accept our position. U is, at present, simply out of the question. That's our deterrence dilemma: we've got no deterrence.

DELOITTE: I take your point. I congratulate you, Dr. Bright, in having gone to the heart of my dilemma. [*He walks up and down, hands behind his back, frowning.*] The position seems to be rather hopeless.

[*Jones looks at the general with concern.*]

BRIGHT: Er...excuse me. [*Jones looks at him sharply.*] May I go through the ways of solving a deterrence dilemma? [*The general stops walking and looks inquiringly at him.*] One: you can, of course, change your position. You must do that if all else fails. Two: you can persuade one of the other parties that the threatened future, or anything they might do in reaction to it, actually makes them worse off than our position.

DELOITTE: But we've just said they prefer it.

BRIGHT: But why do they? Take the Moryan government. They only prefer the threatened future because they're hoping for certain grand strategic reactions, press hysteria, leading to a demand for intervention against ISRA. Can't we persuade them that will never happen?

JONES: We can't get into playing politics.

BRIGHT [*to the general*]: Sir, could we brainstorm this a bit? Without analyzing the grand strategic drama, which we haven't time for, perhaps there's something you could do, make recommendations to your political masters, to get a grand strategic message sent to President Saldin to discourage him from expecting intervention. Or is there some way you could use, or threaten to use, the press conference?

DELOITTE [*staring at him*]: Of course there bloody is! Ray, remember what Intelligence was telling us.

There's evidence the Moryan government may have bombed their own people, that this whole thing was set up by them. That fits in with the way they seem to have suddenly switched their line against peace talks.

JONES: But sir, it was decided not to use that...evidence.

DELOITTE [*belligerently*]:   Decided? So what? [*To Bright*] Let's think about this. If at the press conference I were to present evidence that the government itself is to blame for the bombing...[*Thinking again*] But no. Press reaction might be unfortunate.

BRIGHT: Sir, like most threats, that might be more effective as a threat than it would if carried out. If I may, let me give you a card...

Second Narrated Interlude

[*The narrator steps forward and lifts his hand, freezing the actors in position. He raises the sheet on the flipchart, revealing* table 9.]

NARRATOR [*pointing to rows and columns in* table 9 *to illustrate what he says*]:   Mark now adds the row, "Blame the government," to represent a card held by UNFORMOR. The general then decides to declare his intention to play this card if there's no movement in others' positions, specifically, if the Moryan president refuses to talk. In other words, he makes it part of a new fallback position for UNFORMOR, part of a new threatened future, shown in column t. Brigadier Jones, incidentally, points out that ISRA have in fact blamed the Moryan government for the bombing, which indicates that for UNFORMOR to blame them will be part of ISRA's position, too (column I).

[*Turning to the audience*]: Now, what's going on here, looked at from a technical viewpoint? The general has reacted to his deterrence dilemma, pointed out to him by Mark, by reframing the situation, adding a card to his hand, and inserting that card into the threatened future. This eliminates the dilemma by making the threatened future worse for the Moryan government than his (the general's) position. Notice the emotion with which he carries out this reframing. He becomes angry at the Moryan government. But he doesn't get angry on perceiving his dilemma (then he was merely

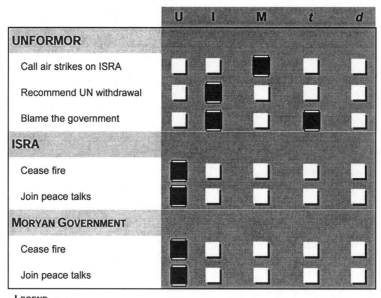

LEGEND
■ means card is played
☐ means card is not played
U is position of UNFORMOR commander
I is position of ISRA
M is position of Moryan government
t is threatened future
d is the default future

Table 9. The general has thought up a
new card, "Blame the government."

dejected, walking up and down, fearing he might have to abandon his position). Nor did he get angry, much earlier, at the Moryan reluctance to talk. He got angry when starting to think of a way to make the threatened future worse for them. Anger is the emotion that accompanies thinking up an adequate threat.

Having thought up the card, the general's anger disappears while he considers how to use it. Does he face any dilemmas in making it credible that he will use this card? Watch…

[*He waves his hand again, bringing the actors back to life.*]

Second Resumption

BRIGHT [*pointing to the new card*]: "Blame the government." If you make that part of the threatened future…

DELOITTE [*thoughtfully*]: …then from President Saldin's point of view, that threatened future would be pretty bad. With us blaming them for the bombing, how could they expect the West to intervene on their side?

BRIGHT: Exactly. Adding this card makes the threatened future worse for them than your position, and so it eliminates your dilemma. Instead, it gives the Moryan government a dilemma, an inducement dilemma, consisting in the fact that they now prefer our position to the threatened future.

DELOITTE: You call that an inducement dilemma?

BRIGHT: Yes, sir. It puts pressure on them to accept our position. It's good for us, bad for them. We, on the other hand, don't prefer their position to the threatened future. We've no inducement dilemma...Let's see, now, what about a threat dilemma...

DELOITTE: Hopefully I will not have to actually play the new card...The threatened future is one that needn't actually be implemented...

BRIGHT: Correct, sir! It's a threat that lies in the background, hopefully making them change, because if there's no change, it represents what's expected to happen...But, if I may, can we look at other dilemmas? Is the new card, "Blame the government," one you'd want to play, sir? For its own sake, I mean, apart from putting pressure on the president?

JONES: No!

DELOITTE: Er... no. I'd obviously prefer not to take sides, upset world opinion, act against the very side that's favored by our own and allied governments...Definitely not preferred.

BRIGHT: Then, sir, we do face a threat dilemma, consisting in the need to make credible a threat we'd rather not carry out. How to overcome it? Well, assuming we don't want to change position, we can use emotion. Getting angry could make them think we'll carry out the threat, even while preferring not to...Better, can we rationalize a change of preferences, that is, give reasons why we would actually prefer to or be forced to carry out the threat? Can we think of any such reasons?

JONES: May I make a suggestion, sir?

DELOITTE: Yes?

JONES [*delicately*]: You might indicate that it would be your duty, if they refuse to talk, to reveal the evidence that puts blame on the government. There is, perhaps, a perception on their part of the British character…

BRIGHT: Right! I see what you mean! A perception that the British can be unreasonably obstinate in pursuit of inexplicable principles! [*Jones nods.*] That way, sir, you could simultaneously admit all the reasons why you'd regret revealing this evidence to the press, while making it clear that you'd feel forced to do so. On principle…

DELOITTE [*decisively*]: Ray, can we get through to Saldin, on the phone?

JONES: Yes sir. The communications room…

DELOITTE: I'm going to do this right now, before the press conference. I can see just how to put this to him.

[*He leaves the room. Bright becomes plunged in thought in front of the flipchart. Jones stands up, walks in front of the table, folds his arms, and gazes thoughtfully at Bright's back.*]

JONES: Dr. Bright, may I ask you something?

BRIGHT [*turning round*]: Of course.

JONES: Is there an…ethical aspect to use of these procedures? It seems somewhat…Machiavellian. Not a straightforward way to deal with people.

BRIGHT: I see. You're right, in a sense, when we're using the procedures in this way, for decision support of one party. Then it's a matter of helping that party to get its own way. We do it by helping it to behave naturally, that is, transmit emotions and arguments appropriate to its situation. That's all right, as far as it goes. But it also means making its behavior consciously appropriate and functional, while leaving all the other parties to stumble around and mess up their chances. So the situation is asymmetrical. You're showing one side how to win.

JONES [*nodding thoughtfully*]: Which is what we want, of course…

BRIGHT: I assume so. I'd like to assume an Allied peace-keeping mission is generally in the right, and ought to get its way. In another kind of situation, we can use the same kind of procedures for mediation support instead. That means working simultaneously with all the parties, getting them to solve dilemmas in such a way as to work out a cooperative solution together…

[*He tails off as the general re-enters the room. Jones smartens up and turns his attention to his commander.*]

DELOITTE: Right. New situation. Assume the president's changed his position.

Third Interlude

[*The narrator steps forward, raises his hand to freeze them in position, and lifts another sheet on the flipchart to reveal* table 10.]

NARRATOR: General Deloitte's blunt speaking to President Saldin has shifted the Moryan government into acceptance of UNFORMOR's position. The general now wants to press home this victory. So Mark

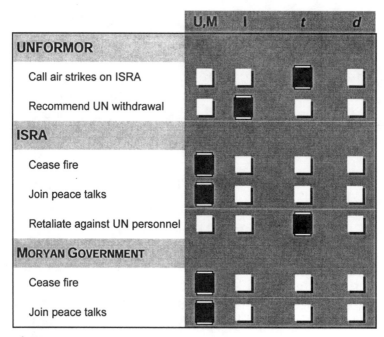

	U,M	I	t	d
UNFORMOR				
Call air strikes on ISRA	□	□	■	□
Recommend UN withdrawal	□	■	□	□
ISRA				
Cease fire	■	□	□	□
Join peace talks	■	□	□	□
Retaliate against UN personnel	□	□	■	□
MORYAN GOVERNMENT				
Cease fire	■	□	□	□
Join peace talks	■	□	□	□

LEGEND
■ means card is played
□ means card is not played
U,M is joint position of UNFORMOR commander
and Moryan government
I is position of ISRA
t is threatened future
d is the default future

Table 10. After the Moryan president has
accepted the UNFORMOR position.

sets up a card-table in which UNFORMOR and the government occupy the same position.

He then asks, what is now UNFORMOR's fallback? Not "Blame the government." That's not necessary any more, now the government has accepted UNFORMOR's position. Does then the fallback consist of doing nothing? If so, the general will face another deterrence dilemma, since ISRA (the only player now openly rejecting his position) will prefer such a threatened future to the general's position.

They reluctantly decide that threatening to call air strikes is all they have to bring pressure on ISRA. This is an action the general would be most unwilling to take. It would jeopardize the relief convoys that depend on ISRA's cooperation to get through. Also, it might cause ISRA to retaliate against UN personnel...but as soon as this possibility is suggested, Mark points out that it is not in the model. It has to be modeled as an important ISRA card. He puts it into the card-table and makes it part of the threatened future, ISRA having more than once indicated that this is what it would do if UNFORMOR were to order air strikes.

The discussion now continues...

[*The narrator waves his hand, bringing the characters back to life, and leaves the stage.*]

Third Resumption

BRIGHT: Okay, so ISRA will threaten us with retaliation. Fair enough. But is their threat credible? Haven't they got a threat dilemma? Does retaliating against UN personnel make the threatened future better for ISRA, in and for itself, disregarding any pressure it puts on us?

DELOITTE [*dubiously*]: Hard to say, in and for itself. That is, forgetting about possible international repercussions.

BRIGHT: Just a moment, sir, we shouldn't forget about them. They'll be part of the grand strategic game, which, remember, we have not analyzed, but have to consider when analyzing this one. International repercussions...Let's see...you mean that if they retaliate against UN personnel, it's likely the UN and the powers behind it would demand intervention against ISRA?

DELOITTE: Precisely. Don't you agree, Ray?

JONES: I do, sir.

BRIGHT: Okay, then if we take grand strategic repercussions into account, ISRA should have a threat dilemma in threatening retaliation. If they realize it. I mean, if they realize the likelihood of repercussions, they should prefer not to retaliate against UN personnel...If they don't realize it, perhaps we should point it out to them! Which suggests something else...We might at the same time get rid of our own threat dilemma, I mean the fact that they'll think us reluctant to carry out air strikes.

DELOITTE: There you've lost me, I'm afraid. Go over that slowly.

BRIGHT: Yes. I mean this, sir. We've established, I think, that heavy, sustained air strikes against ISRA positions would solve our deterrence dilemma. That is, rather than suffer that, they'd prefer to accept our position, which is a cease-fire and peace talks.

JONES: ...Although we've said their acceptance mightn't be very trustworthy...

BRIGHT: Right, agreed. I'm not forgetting that. That is to say that accepting our position might give ISRA (and the government, too, come to that) a cooperation dilemma, in that they'd be accepting a position it pays them not to honestly implement. Likewise it'd give us a trust dilemma, we'd have got them to agree to something we can't trust them to do. But those are dilemmas to be tackled once we get overt acceptance of our position. Getting overt acceptance is the first step. That's what we're discussing now.

DELOITTE: One step at a time.

BRIGHT: Exactly, sir. Now, to get overt acceptance we must make it the best alternative for them that they can see. That means, first, making the alternative, the threatened future, worse for them, that is, solving our deterrence dilemma. Air strikes do that. Correct?

DELOITTE: I believe so.

BRIGHT: Second, it means making air strikes credible. That means getting rid of our threat dilemma. Making them believe we prefer to or will be forced to carry out air strikes, if they don't accept our position.

DELOITTE [*concentrating on this*]: …Even though I don't, in fact, prefer to carry them out, because, as we've said, it'll mean an end to relief convoys plus a likelihood of retaliation.

BRIGHT: Correct, sir, but we have to get rid of that impression in the minds of ISRA. How do we do that? Well, we could emphasize to them the certainty of international repercussions if they retaliate. If they become convinced of that, their retaliation may become unlikely, eliminating that particular reason for our dislike of air strikes.

DELOITTE: Okay. But how do we eliminate our other reason for disliking air strikes? The fact that it will disrupt relief convoys?

BRIGHT: Er, maybe I'm confused here. I had in mind, again, emphasizing international repercussions as the reason why...

DELOITTE [*exuberantly, completing his sentence for him*]: ...as the reason why I'll be forced to call air strikes if they refuse a cease-fire, regardless of whether I like air strikes or not! Jolly good! This works, Ray. I'd been planning, in the usual kind of way, to dampen down the tendency toward escalation. Now I'm seeing that tendency as my strength. I mean to use it like a following wind. Unless we get a cease-fire and talks pronto, then the pro-intervention movement in the West will force my hand.

JONES [*dubiously*]: There certainly is such a movement, sir, pressing for intervention against ISRA.

DELOITTE [*walking up and down with hands behind his back*]: Indeed there is. What this has made me see [*waving his hand at the flipchart*] is that to avoid intervention, I shouldn't seem to oppose it. Those strong pro-intervention forces are what give my threat, the threat of air strikes, credibility. And without such credibility, ISRA will not agree to a cease-fire, things will continue to get worse, until eventually we're forced to intervene on a larger scale and with less chance of success.

BRIGHT [*hurriedly*]: Of course, what you say at the press conference and through other channels may still fail to make air strikes credible, sir. ISRA may still think they can call our bluff, given our past history of threats not followed up. In which case, credibility may require starting to carry out air strikes, in other words, shifting the default future to one that contains air strikes.

DELOITTE [*after walking up and down silently for a while*]: Right. So I may have to start air strikes, while making it clear that if they agree to a cease-fire and talks, air strikes will cease. We should, therefore, start withdrawing our people; it'll take a while anyhow to locate targets properly. [*He stops, looks at his watch and addresses Mark.*] Dr. Bright, this has been useful. I'm off to the press conference now with a clear overall plan to pursue. I want you to continue this analysis. Brigadier Jones will make sure you get the help you need, including access to information—the, er, assumptions you need to put into your model. What will you look at next?

BRIGHT: I'd like to explicitly model both the grand strategic as well as the military strategic games, sir. I think we need to be clear as to how they work, how they impact on your own, operational level. [*The general nods.*] Then I'll need to model some of the confrontations going on at tactical level. We should plan to send a single, unified message at all levels. That way we can bring it about that ISRA and the Moryan government are being pushed in the same direction by pressures from their own grass-roots as we are pushing them at the higher level. That'll be important for making air strikes credible. It'll be still more important when it comes to making sure they keep to our position, that they actually implement it, as distinct from merely agreeing to it to keep us off their backs.

DELOITTE [*who has been listening intently*]: Right. Carry on. [*Turning to Jones*] Ray...

JONES [*opening the door for him*]: Yes sir. Your car must be waiting. I'll see Dr. Bright gets what he needs.

[*They leave together. Mark Bright collapses on a chair and grins at the audience.*]

BRIGHT: Whew! Well, that seemed to go all right. What did you think?

SUMMARY OF CHAPTER 6

The chapter dramatizes the situation of a commander who, knowing nothing of confrontation analysis, formulates and begins to implement a confrontation strategy with the help of an analyst who happens to be on his staff. This is not how it would ideally be done, but serves to illustrate the concept of a confrontation strategy before a discussion of its principles in the next chapter.

General Deloitte has been tasked to get two warring parties to cease fire and start peace talks. He is not to use force. He has to address a press conference in an hour, and is anxious to formulate a plan. The analyst called in to help him draws up a card-table on a flipchart, points out that he has a deterrence dilemma and suggests directions in which he might look to try to eliminate it. This prompts the general to think of a way forward by adding a new card to his threat against one of the parties. After discussion of how to remove his resultant inducement and threat dilemmas, he implements this plan through a phone call, which brings the party concerned into apparent compliance with his mission objectives (i.e., they now say they accept his position). Of course, this party may have a cooperation dilemma, so that the general may still have a trust dilemma; this, however, can be dealt with later.

A deterrence dilemma remains in relation to the other party, which, they decide, can only be eliminated by threatening to use force. A threat of retaliation against UN personnel can, they decide, itself be deterred. The general's own threat and inducement dilemmas can be removed in the eyes of the other party. The general now has a clear plan upon which to proceed in dealing with the press conference. He leaves after asking the analyst to build models of the grand strategic, military strategic, and tactical levels.

CHAPTER 7

FORMULATING A CONFRONTATION STRATEGY

The last chapter showed a confrontation strategy in use, although not as it should be used, not implemented in an organized way by personnel trained in how to do it. We saw it adopted in an ad hoc manner by a commander relying on a single expert who happened to be on his staff.

Nevertheless, we now have an example to fix on. Next is a discussion of the principles that should underlie the formulation and implementation of a confrontation strategy.

WHAT IS A CONFRONTATION STRATEGY?

We have said that Operations Other Than War (OOTW), and peace operations in particular, can be viewed as sequences of linked confrontations; however, the confrontations involved are not all on one level.

To see what a confrontation strategy is we need to see how it is implemented at various levels. We begin by discussing its implementation at grand strategic, military strategic, operational, and tactical levels of command. There are many other levels within each of these four; however, discussion of these broadly defined levels will present the general idea.

Then follows a discussion of what confrontation strategy conducted on these four levels consists of. To illustrate, we use two examples: the roadblock-removal in tables 2 and 3 found in chapter 2 and General Deloitte's strategy in chapter 6.

How Different Levels of Command Implement a Confrontation Strategy

First, the grand strategic or political level is that at which, in the play, Western public opinion was seen to be putting pressure on Western governments to stop ISRA from massacring people. A confrontation strategy at this level is primarily the responsibility of politicians. They are dealing with the public, with each other, and with foreign governments. As in all confrontations, they are trying to orchestrate the emotions and arguments provoked by various dilemmas to get everyone to play their tune.

Some of the cards politicians can play are military; therefore, the military's role is to advise them on the practicability and likely consequences of playing one or another military card. The military also needs to understand the grand strategic confrontations that politicians are involved in to understand the intent of their superior and their superior's superior.

At the military strategic level, senior military staff maintain and deploy a nation's total defense forces in pursuit of national political objectives. Having received directions from the grand strategic level, this level must issue strategic directives to operational-level commanders. In our terms, these directives empower a commander, in the case of an OOTW, to implement a confrontation strategy against certain other players in pursuit of certain objectives. They also give him certain types of cards he can play; therefore, it is necessary for staff at the grand strategic level to have a broad understanding of the confrontation strategy required of their OOTW commanders to pursue. In addition, they are involved in various, hopefully cooperative, confrontations at their own level (e.g., among themselves or with the military staff of other countries with whom they are required to act in coalition).

The next level, the operational level, is the level at which General Deloitte is operating in the play. Having received his strategic directive from the British national defense staff, his task is to play his cards in such a way as to achieve his mission objectives. To do so, he must confront various local parties and try to get them to comply with what he wants. In the play he is planning to do this and, at one point, actually leaves the room to do it. In this, he is implementing a confrontation strategy.

At the operational level General Deloitte, like the staff at grand strategic level, will conduct various other confrontations, hopefully cooperative, such as horizontal ones with other units in a joint or combined force and internal ones with his own staff. In the play we see him in a sequence of such confrontations with his chief of staff and Mark Bright. (These personal confrontations are not analyzed, just acted out.)

Finally, the general's confrontation strategy needs to be analyzed and broken down into strategies that can be carried out by units reporting to him at the tactical level. The significant thing about the tactical level is, in general, that this is where things, physical things, actually get done.

What things? In the case of battle-fighting, tactical-level units carry out and support destructive military operations. In the case of an OOTW, destructive military operations are not necessarily required. When they are, their primary function often is to send a message rather than to cause physical destruction. Non-destructive operations, such as assisting refugees and helping with disaster relief, may be required. Even then, the ultimate objective is to get others to carry on such operations, perhaps after receiving initial or interim help, rather than for the military to do so indefinitely.

Generally it is true in an OOTW that the objective is not so much to do physical things as to get others, those on the spot who will remain after we have gone, to do or refrain from doing things. (Operations such as disaster relief may seem to be an exception. Here there are certainly physical things to be accomplished. Yet even here the need to obtain cooperation with other agencies and local bodies is often paramount.)

In a sense the above is true of all military operations. Clausewitz himself says that after having "render [-ed the enemy] incapable of further resistance" we "compel [him] to fulfill our will" (1968, 1st edition 1832; p. 101). In Clausewitz's traditional view, the military task, as such, was to "render him incapable of further resistance." After that was done, politicians took over to dictate to him our will. In an OOTW, by contrast, the military's task is to induce or compel others to fulfill our will, if possible without first rendering them incapable of further resistance.

This is the objective and essential principle of a confrontation strategy. In relation to what is done at the tactical level, the point is that while we are conducting a confrontation, we are not essentially doing anything except communicating. Here "essentially" means that we may be doing other things, but all such other things will have communication as at least one of their functions, and it is this function that our confrontation strategy is concerned with.

In devolving a confrontation strategy to the doing, or tactical level, we again must devolve, not a number of physical tasks as such, but a number of cards to be used in confronting others and getting them to comply with our objectives. The essential difference, nevertheless, between the operational and tactical levels is that tactical level confrontations generally are conducted at the local, grassroots level.

In devolving an operational level strategy, the essential principle is as follows: To plan tactical-level confrontations to change others parties' foot-soldiers in such a way that the message they send up to their theater-level commanders reinforces the message we are sending to those commanders at the operational level. This is the essence of how to devolve a confrontation strategy to the tactical level.

Linked Confrontations in a Confrontational Campaign

We see that an OOTW campaign is not a single confrontation, but a number of confrontations at different levels linked to each other. Following is a list of some of those confrontations:

- An operational commander's mission is defined for him at the military strategic level after politicians in a confrontation at the grand strategic level have decided to play certain military cards. It is the political confrontation at the grand strategic level that defines his superior's intent and his superior's superior's intent.

- The operational commander is then involved in various (hopefully cooperative) horizontal confrontations with other units in a combined or joint force; he must resolve these to properly deploy his military assets.

- His main confrontation is with other parties that must be brought into compliance with his mission objectives.

- Finally, the operational commander must devolve his main confrontation strategy to the tactical units he commands. These then face numerous local confrontations at the tactical level.

We are now close to what a confrontation strategy is. But first we need to say how a theater commander might, in a simple, step-by-step process, form and implement a confrontation strategy for his main confrontation, that in which he must get other parties to comply with his mission objectives. Links with other confrontations will be discussed where relevant to this description of a confrontation strategy in action. After describing a confrontation strategy in action, we can say in general what a confrontation strategy is. Following is a step-by-step description.

A Confrontation Strategy in Action, Step by Step

In **step 1**, a commander builds a model of his confrontation and identifies within it the dilemmas faced by all parties. These include his own and others' weak and strong points (i.e., the points at which he and others are open to being changed).

In the play, Mark Bright helps General Deloitte and his chief of staff build a model, which is essentially built and owned by the general, with all substantive assumptions belonging to the general, while Mark provides expert advice and guidance on what a confrontation model is. Mark might argue for or question a certain assumption on the basis of his knowledge of the theoretical question being answered by it. He does this quite often in the play. After he is sure the general knows the meaning of a question, he always accepts the general's answer. After arriving at a first model, Mark helps the general identify that he faces a deterrence dilemma. He does not, as perhaps he should, go on to check that he has identified all dilemmas; instead, he immediately asks the general to consider what to do about this one. He is taking short cuts based on his knowledge of model building. After all, the general asked him to do what he can in an hour.

In the model shown in table 2, the commander might identify himself as facing a trust dilemma and the ethnic militia as facing a cooperation dilemma. On the other hand, the commander does not have a threat dilemma. Rather than allow the militia to keep roadblocks in place, he would prefer to remove them forcibly. His first preference would be that the militia should remove the roadblocks themselves.

In **step 2**, after the commander has identified the dilemmas in his model, he chooses a subset of dilemmas he will attempt to eliminate and simultaneously establishes the means by which he will attempt such elimination. He must use his judgment and knowledge of the situation to estimate how much friction would have to be overcome in bringing about various dilemma-eliminating changes.

General Deloitte decides to eliminate his deterrence dilemma by introducing a new card, "Blame the government," and making it part of his fallback position. Realizing that this will give him a threat dilemma, he next decides to make the Moryan president believe he will feel bound to play this card if the president does not back a cease fire. He spends some time discussing the feasibility of these changes (i.e., how much friction he must overcome in introducing this card and making the president believe he will play it). Notice that the two dilemmas he is eliminating are connected: the second is brought about by the elimination of the first. Notice too that there is a strong connection with the grand strategic drama. Bright does not analyze it, he does not have time, but the connection is as follows. The Moryan president is trying to appeal over the general's head to international opinion and Western governments to get them to send directives to the general that will change his preferences (i.e., make him prefer the

Moryan position (M) in table 8 to the threatened future *t*. The general will have to prefer it, the president reckons, if he is directed to implement it.) In this way, the president is trying to overcome a deterrence dilemma, consisting in the fact that the general is under no pressure to accept the president's position. The president is trying to overcome this dilemma by making moves in the grand strategic confrontation that will, he hopes, affect the present drama through linkages between the two confrontations.

In table 2, the commander finds he has a problem. The militia has signed the agreement but is continuing to maintain roadblocks. When questioned, the militia leadership denies that there are roadblocks or blames them on local groups. Meanwhile the militia points out that the commander himself is violating the agreement by withholding aid and support. Shall the commander give up on the agreement and move to the threatened future *t* (i.e., forcibly remove all roadblocks)? (This was IFOR policy after the Dayton agreement.) Or shall he try to change the militia's preference for unofficially keeping roadblocks in place? He decides to try the latter. He devolves his strategy to the tactical level, directing local commanders such as the one in table 3 to persuade local militia to remove roadblocks. He tells the militia leadership that this is what he is doing. He is, he says, helping them to make their policy effective with local groups so that aid and support can start to flow. Simultaneously, he starts a campaign on local television telling local militia groups what is the official policy of their leadership. He reasons that if local commanders, thus supported, succeed in their local confrontations, then the leadership of the ethnic militia will change its preferences and abandon its unofficial policy of maintaining roadblocks. (Note that this is a link that he sees between the two levels.) He will then let them have the promised aid and support.

Step 3 is to implement the planned dilemma-eliminating measures. Note that by definition no one can know, when entering a moment of truth, how they will be changed by it. This applies less to the commander, who is deliberately instigating a moment of truth to cause specific dilemma-eliminating changes, than to those he is interacting with, who are, in comparison with him, unprepared for the pressures they are about to face; nevertheless it applies in some degree to both sides.

In the play, General Deloitte gets on the phone to President Saldin and lets him know he had better come to the cease-fire talks and back a cease fire, or he will be publicly blamed for being responsible for the massacre. This, he must see, will undermine his efforts to enlist Western support. President Saldin now faces a moment of truth.

The UN commander in table 2 sends directives to his local commanders, describing his confrontation strategy and how they can contribute to it, giving them appropriate cards to play but leaving the details of their individual strategies to their initiative as commanders. If they have questions or suggestions, such as additional cards they might want to have in their hand, they discuss them with him. At the same time he informs the militia leadership of what he is doing, presenting it as a way of helping them to exert control over local groups and holding out to them the prospect of aid and support. Finally, he launches a television campaign while his local commanders confront local groups manning roadblocks. In this way, the militia leadership is brought to a moment of truth.

Step 4, following the moment of truth, at which changes were meant to take place, is a return to step 1 to build a model representing the new, changed situation. If there is now a satisfactory solution, with all parties taking the same position and trusting each other to carry it out, then the commander has achieved his mission objectives (perhaps in a modified form). If not, dilemmas remain, which the commander identifies at this point. He then moves on to step 2, and so on.

In the play we saw how the commander, having brought the Moryan government into compliance with his objectives, models the new situation and works out how to do the same to ISRA.

Suppose, in table 2, that the UN commander's strategy succeeds in getting the militia to comply with the removal of roadblocks. There are now no more dilemmas. The problem is resolved within this model. Other, more detailed models, or models focusing on other issues, would no doubt reveal difficulties. In addition, implementation of the agreement may meet unforeseen problems. For example, the commander may find that the aid and support he promised to the militia, and which he has held up by refusing to certify that the first stage of pacification has been reached, is now held up for other reasons. To get around this problem, he must resolve a horizontal confrontation with the high representative responsible for aid coordination.

Note: A confrontation between two parties is vertical if they are related as superior and subordinate in the same organization. This means that one party generally has a card consisting of formally issuing orders or directives to the other, although often this does not settle the matter; if the other does not agree, it may be able to ignore or re-interpret directives given to it, or appeal against them. The distinction between vertical and horizontal relations becomes fuzzy when, as is often the case in peace operations, a commander is responsible both to his national headquarters and to an international organization. For the sake of our example, we are assuming that the high representative has no authority over the UN commander (as in the case of the Dayton agreement, where the UN high representative was responsible only for coordinating civilian aspects of the agreement). Therefore, negotiations between the UN commander and the high representative in our example are horizontal. Neither has authority over the other.

The Confrontation Strategy Itself

The above step-by-step description of the implementation of a confrontation strategy is not yet a description of the strategy itself, but it enables us to give the following definition. A confrontation strategy consists of a plan for cycling through the above four steps until a resolution of the confrontation is reached; therefore, it consists of a sequence of dilemma-revealing models of characters' positions such that:

- A plan for dilemma-elimination is associated with each model except the last.

- The next model in the sequence is created by implementing the plan on each model except the last.

- The last model has no dilemmas.

This is our suggested characterization of a confrontation strategy. To understand it correctly, several points must be kept in mind.

First, each plan for dilemma-elimination associated with a model in the sequence is a plan for bringing about a moment of truth, at which changes may take place. We have said that on entering a moment of truth no one can tell exactly how it may change them. This, as well as the general fact that the future tends to hold surprises, makes it an essential characteristic of a confrontation strategy that it should be looked at anew, and reformulated if necessary, at each stage, after each moment of truth.

Even though it is constantly reformulated, it is nevertheless desirable for a whole strategy to exist, so that assets can be mobilized and relevant information collected to fulfill it.

Second, it is important to realize that a plan for dilemma-elimination associated with a particular model in the sequence may be linked to other, related confrontations. We have looked at some examples of such linkages between the particular confrontation a commander is involved in and other confrontations.

For example, the confrontation between General Deloitte and President Saldin was linked to the grand strategic confrontation between the West, the Moryan government, ISRA, and other countries (e.g., the Arab nations, France, the bordering state of Pelugya). Part of this linkage consisted in the fact that since General Deloitte's job was to fulfill his mission, his preferences were largely linked to the policy decisions of the West. If Western politicians wanted him to side militarily with the Moryan government, that likely would become his preference. The result was that to eliminate a dilemma in his confrontation with the general, President Saldin could try to alter the latter's preferences by means of moves in the grand strategic confrontation he was conducting with the West.

The confrontation in table 2 between the UN commander and the ethnic militia leadership was similarly linked to the confrontations between local commanders and local militia groups; that is, decisions by a superior could decide the preferences of a subordinate. However, interestingly, the opposite was also true in the case of the ethnic militia. The commander's plan for eliminating a trust dilemma depended on lower-level militia attitudes influencing the preferences of their leadership.

After apparently resolving his confrontation with the ethnic militia, we saw the UN commander finding himself unable to implement the agreement reached because of problems with a horizontally related partner, the high representative responsible for coordinating aid. Technically this meant the commander had to resolve a horizontal confrontation to eliminate a cooperation dilemma.

APPLYING THE PRINCIPLES OF WAR TO CONFRONTATIONS

It is clear that a confrontation strategy is different from a strategy for warfighting, yet some of the same principles apply; therefore, it may be useful to ask which generally known principles of war continue to apply to confrontations. This may help to bring out what is different about OOTW. We are not attempting any more than this; we do not aim to establish a list of principles of OOTW.

Seven Principles of War

Alberts and Hayes (1995, pp. 28-37) ask how far the principles of war apply to peace operations. They survey seven principles of war and conclude that *simplicity* and *unity of purpose* (rather than of command) remain important. Unity of command, they point out, is often unachievable when there is a need to coordinate strategy between different national contingents; it has to be replaced by the attempt to achieve unity of purpose. Achieving this also helps to *focus on the objective*. Consequently they suggest that unity of purpose can replace this principle also.

Thus, Alberts and Hayes endorse simplicity as a principle for peace operations and subsume the two principles, unity of command and focus on the objective, under the one principle, unity of purpose. They suggest adding *consensus planning*, *adaptive control*, and *transparency of operations* as principles to be followed in peace operations. Four other principles that apply in warfighting, *taking the offensive*, *concentration of superior force*, taking the enemy by *surprise*, and *maintaining security*, are, they consider, less applicable and may even do harm.

This is certainly right if we look at the physical operations involved, as we do when considering warfighting. We have said, for example, that a physical operation such as a bombardment or a troop movement whose primary purpose is to send a message cannot sensibly be kept secret. Thus the principles of surprise and security, applied in ways that would make sense if the primary purpose were destruction of the enemy, can do harm if applied in OOTW.

The Contrast Between Warfighting and Confrontation

The contrast involved here is important; it perhaps needs re-emphasizing. If we are bombarding the enemy to force negotiation, our bombardment is actually a message to the effect: "Look what will happen to you if you don't come to the bargaining table." Its physical effects in actually destroying the enemy's assets (as distinct from communicating our determination to do so) are a disadvantage to us as well as to the enemy, because the more assets we have already destroyed, the less remain to be destroyed, and the less significant is our threat to continue destroying them.

Counter-insurgency forces, for example, by destroying the homes and families of guerrilla-fighters, often create large numbers of homeless, bereaved fighters with nothing left to lose and much to revenge.

Consequently, to send the message, "look what will happen to you," the ideal is to combine a maximum of shock and awe with a minimum of actual destruction. Keeping secret the destruction you have wrought is counter-productive. By contrast, if the intention is not to send a message but to actually destroy the enemy's fighting potential, then the more secure he feels and the less he knows about what is happening to him the better, because his ignorance and contentment reduce the effectiveness of his countermeasures and maximize the damage done to him.

On the other hand it may be a sensible strategy, as we have seen, to interrupt negotiations to weaken another's bargaining position by physically taking certain cards out of the other party's hand. To complicate matters, such physical operations may be combined with message-sending. For example, operations such as interdicting terrorists' communications or destroying their arms dumps may reduce their ability to harm us, and thus either take cards from their hand or reduce the effectiveness of certain cards they might play. Such operations obviously require secrecy and surprise at the tactical level, but if carried out while we are conducting a confrontation with terrorist leaders, they may serve two functions: to send a message ("This is what you can expect if you don't concede") as well as to diminish the effectiveness of terrorist action. The message may even be, "You can expect your effectiveness to keep diminishing as long as you hold out."

An operation that has dual functions, message-sending combined with physical effects, may need a balance of secrecy and openness. It remains true that a message as such needs to be clearly received and correctly interpreted if it is to do its job. To destroy terrorists' effectiveness will not work well as a message if the terrorists think they have suffered less loss of effectiveness than they actually have, or mistakenly believe that we mean our strikes against them to be our last. Yet these mistaken beliefs actually would benefit us if our actions were not meant to send a message, but were carried out purely to diminish the terrorists' physical effectiveness.

The contrast between something done for its actual effect and the same or a similar thing done to send a message is striking. We have said that a confrontation strategy (the multilevel, coordinated object described in the last section) is not essentially a matter of conducting physical operations, but of sending messages. It is not surprising that we get somewhat different results when we look at how the Alberts-Hayes principles of war apply to message-sending rather than when they are applied to physical activities as such.

Applying the Principles to a Confrontation Strategy

Consider now the seven principles in turn.

Simplicity is of utmost importance, essentially for the reasons given by Alberts and Hayes, but we can add some more. We have seen that an operational commander's confrontation strategy needs to be understood and implemented in a coordinated way by tactical echelons below him and other task-force units horizontally related to him. He also needs to be able to explain it clearly to military strategic and grand strategic levels above him to enlist their support in pressuring parties from above or expanding the cards available to him, when necessary. This is the confrontation-strategy need for simplicity. Simplicity is also necessary because the common reference frame through which the commander communicates his position to other parties must be simple; therefore, his confrontation strategy can be simply expressed in terms of it. The commander in table 2, for example, might explain his strategy by presenting and explaining tables 2 and 3.

Unity of purpose, including, as Alberts and Hayes suggest, *focus on the objective,* also magnifies the effectiveness of a confrontation strategy. Suppose a commander is pressuring another party to take a certain action, say to cease fire, withdraw armaments, or permit the passage of refugees, and to this end is using certain implied threats and promises. We have seen that he is more likely to succeed if the other party's grassroots organization, superiors, and allies are receiving versions of the same message, and so are inclined to tell it to make the concession. Such a unanimous chorus is achieved only by getting unity of purpose behind a message dictated by an agreed confrontation strategy. We have also seen that obtaining agreement between different coalition partners on a common confrontation strategy is itself a task for a (separate, hopefully cooperative) confrontation strategy.

Taking the offensive is shown by Alberts and Hayes to be an inappropriate principle for peace operations, if we think in terms of a physical offensive; however, as applied to a confrontation strategy, we might reinterpret it to mean taking the initiative in forcing a moment of truth so that a certain set of dilemma eliminations, planned by us in advance, can occur.

This is to take and hold the initiative in terms of the argument. It is analogous to the battle-fighting concept of forcing the enemy to respond to us, rather than vice versa. To avoid confusion, it might be better to avoid metaphor and call it taking the initiative rather than taking the offensive.

Is *concentration of superior force* another principle we can usefully reinterpret in application to a confrontation strategy? "Superior force" might be taken to mean an array of superior arguments for use at a preplanned moment of truth, provided we understand that the arguments we have in mind need not be very intellectual, but may include points such as "We'll bomb you to bits if you don't." Here "argument" may be the wrong word. "Force" may also be misleading. "Sanctions" is perhaps the right term, because it may refer to positive enticements such as offers of aid as well as negative ones. This suggests that "concentration of superior sanctions" might be the right name for this principle.

Concentration of sanctions may be unnecessary because of what appears to be an essential difference between confrontations and battle-fighting. In a battle, assets used in one place cannot at the same time be used in another; therefore, they must be concentrated before an assault. By contrast, negative assets used in a confrontation, that is, the cards you can play to out-escalate another, are not necessarily used. The hope is that you will merely have to threaten their use. The same negative assets can therefore be used in different places at once. Following are two examples:

- All the commanders confronting local militia in our roadblocks example can simultaneously threaten to call down air strikes. They will not need that many planes.

• The British Empire, while it lasted, was held down with a small number of troops. Each potential rebel feared that those troops would be used against him.

Positive assets (sanctions that are promises rather than threats) are different. They will be used up in a successful operation because using them up is, by definition, the objective. Their quantity must be sufficient to cover the different points where they are used.

In sum, it seems that assemblage of superior sanctions may be the principle that should replace concentration of superior force.

Surprise is another principle inapplicable to OOTW when we consider only physical operations. Given the need to build trust and verify agreed physical conditions (such as separation of forces or disarmament), peace forces should not create uncertainty by behaving in unexpected ways (Alberts and Hayes, 1995, p. 30). Indeed, this follows from our assumption that actions in a confrontation are seen primarily as communications. Our argument is as follows. One of the characteristics of a surprise action is that it is uninterpreted. You wonder what it means. That defeats the purpose of an action that is supposed to send a message. The whole point of that is for you to know what it means. Prior notice tells other parties what an action means.

The principle of surprise, however, may be applicable to the content (i.e., meaning) of the messages you send in a confrontation, as distinct from the form (the action itself). Producing a surprising new argument, offer, or threat, as General Deloitte in the play does in threatening to blame the president for the massacre, makes it hard for the other to produce a convincing defense, because they have not been given time. Lawyers in television courtroom dramas win this way. In the other direction, a surprising concession or compliment, even if insignificant, may cause favorable changes in others' perceptions of your motives just because it is surprising. A surprise draws attention to itself. The principle of surprise, understood in this way, can be used alongside taking the initiative and assembling superior sanctions. They are mutually reinforcing methods of obtaining other parties' willing compliance.

Finally, Alberts and Hayes cast doubt on the principle of *security*, pointing out how over-emphasis on physical security, e.g., ensuring minimal risk to troops, may reduce the effectiveness of peace operations. This follows from asking, "What message does this action send?" Strict security precautions say, "I don't trust you." This is not compatible with "Trust me." On the other hand, it is quite compatible with the message, "I'm ready to smash you," which may, in other circumstances, be the intended message. In relation to physical security, all that confrontation analysis can say in general is what it says in relation to other physical activities: "Pay attention to the message you want to convey."

On the other hand, surprise requires security, and we have said that surprise may be advantageous in itself and help us to take the initiative and assemble superior sanctions; however, this applies to the content of messages, not to physical actions. To ensure surprise in our confrontation strategy, and also to preserve simplicity and avoid misunderstandings, it is advisable to preserve security in the process of confrontation-strategy formation; however, in implementing a confrontation strategy, disclosure of our strategy is usually the best tactic because it reinforces the message being sent and helps to avoid misinterpretation. Similar reasons are given by Alberts and Hayes for their principle of transparency of operations. They are the reasons why, in our example, the UN commander decides to inform the ethnic militia leadership of his devolved strategy of confronting local militia roadblocks.

This seems to leave us, in this area, with a rather awkward hybrid slogan: Security of planning plus transparency of operations.

Summing up, the principles of war apply rather badly to OOTW considered as physical activities; however, many of them seem to apply better to the formation and implementation of a confrontation strategy, which is primarily a matter of sending messages.

SUMMARY OF CHAPTER 7

A confrontation strategy for a particular confrontation consists of a plan for sequentially eliminating dilemmas, each in a specific way, so that the end result is full compliance on the part of all parties with our mission objectives. A strategy for a peace campaign as a whole is a plan for winning in this manner a linked set of confrontations.

A strategy needs to be adaptive to changing circumstances and new information, such as the fact that a particular confrontation is not resolved in the manner expected. At the same time it needs to be sufficiently robust to give continuing guidance throughout the inevitable fluctuations of expectations that occur in the course of a campaign.

For these reasons, a strategy must have a clearly stated objective; be simple; and be carried out with unity of purpose. As has been pointed out, other principles of war, such as surprise, concentration of force, taking the offensive, and security do not apply to peace operations in the same way as they do to warfighting; however, applying superior force and taking the initiative are applicable to confrontation strategy at the level of persuasion, as distinct from physical operations.

A strategy needs to be understood and implemented at various vertically linked levels of command and by various horizontally linked components and coalition forces. We must understand the manner in which different confrontations are linked to formulate a confrontation strategy. A confrontation is linked to others that its conditions and objectives affect or are affected by. It is linked to higher-level confrontations that set the scene for it and to lower-level confrontations that its Implementation phase sets the scene for; also, a confrontation is linked to the more detailed confrontations that take place in its Resolution and Conflict phases.

Confrontations are linked together strategically if the positions and the sequence of dilemma-eliminations adopted in one confrontation are linked to those adopted in others. In the removing-roadblocks example, a confrontation strategy worked out by the theater commander could be devolved to the level of the local commander by linking their confrontations. Similarly, the commander's confrontation with non-compliant parties at theater level should be linked to those of horizontally related characters such as other component and coalition forces. Achieving confrontation strategies that are coordinated between horizontally and vertically linked partners may require consensus planning and negotiation of strategies. Above all, it requires that the commander's strategy be simple and his objectives clear and agreed.

CHAPTER 8

ANALYSIS OF THE BOMBARDMENT OF SARAJEVO, FEBRUARY 1994

On February 5, 1994, a bomb exploded in Sarajevo's market square. It killed 69 people and left 200 wounded. Blame naturally fell on the Bosnian Serbs, who at the time were bombarding the city with mortars placed on the surrounding hills. They at once denied responsibility and blamed the Muslim defenders of the city. Nevertheless, there was an international outcry against the Serbs. As a result, the British commander of the United Nations Protection Force (UNPROFOR) in Bosnia, LTG Sir Michael Rose, found himself tasked with getting the Serbs to cease fire and withdraw their weapons in order to fend off NATO air strikes on their positions.

In this chapter we examine this crisis in some detail, going through the most complex confrontational analysis of a real-world Operation Other Than War (OOTW) that has been done so far. As we do so, we will discuss some of the problems of real-world confrontation analysis.

Sources for the Analysis

Research for the analysis of Sarajevo, February 1994, was carried out by the author in early 1997 under a contract issued by the UK Defense Evaluation and Research Agency (DERA) for a report on the use of drama theory in operational analysis (DERA, 1997). It is reported here by permission of DERA.

The research included interviews with a number of British officers who had served in Bosnia; however, they were not interviewed for their involvement in the particular confrontation analyzed, the crisis over the Bosnian Serb bombardment of Sarajevo in February 1994. All factual assumptions made in the analysis are drawn from published sources. If they are inaccurate, the fault lies either with those sources or with our misinterpretation of them, and we can only hope that those in the know, in particular General Rose and his staff, will make allowances accordingly.

Even if we had been able to interview actual participants, they would not have been the best source for assumptions to go into our analysis. Memories are biased, rationalized reconstructions of what was felt and thought at the time. We tend to think that what did not happen, could not have happened. We do not think so at the time. The best source of assumptions for confrontation analysis is the judgments of those involved while the confrontation is going on. This is how assumptions are made in the fictitious case described in the Frontline play in chapter 6.

Incidentally, the reader will notice that the play, although fictitious, has a plot that is based on the crisis in Sarajevo in February 1994. The play was written by taking that real-life scenario, simplifying it, equipping it with imaginary characters, transferring it to a non-existent North African location, and imagining that it was dealt with by a commander who decides to use confrontation analysis.

Conclusion of the DERA Report

The conclusion to keep in mind and examine critically as you read this chapter and later ones is as follows: If the UNPROFOR commander in Sarajevo had used confrontation analysis in his efforts to get the two warring parties to cease fire, he might have achieved his mission objectives more quickly and satisfactorily.

In the DERA report, the conclusion is spelled out as follows:

- The first stage, bringing the Bosnian government into apparent compliance, might have been achieved in a way that was planned and intended beforehand, without depending on emotional, spontaneous, unplanned reactions of the commander. Emotions might have been used, but in a rationally planned way.

- The second stage, bringing the Bosnian Serbs into apparent compliance, might have been achieved sooner by the UNPROFOR commander on the operational level instead of at the last moment, after many concessions had been made, as a result of developments on the grand strategic level.

- Even if the result had differed little from the actual one, it could have been achieved by the commander in a more purposeful way because from the beginning he would have been able to seize and maintain the initiative.

PUTTING PRESSURE ON THE BOSNIAN GOVERNMENT

When the bomb exploded in a square crowded with shoppers, the international outcry brought to a head a growing demand for intervention to stop the Serb bombardment of Sarajevo. The UNPROFOR commander actually was engaged at the time in negotiations to this end, tackling the problem that while the Serbs were predominant in heavy weapons, they were weaker in infantry than the (mainly Muslim) Bosnian government forces defending the city, who were ready to take the offensive if Serb actions were halted.

At this time, the UNPROFOR commander conceived his mission as primarily humanitarian (delivery of aid, prisoner exchanges). After the bomb fell, he went to Belgrade, met with the overall UNPROFOR command and the UN special attaché, and came back with a mission to get the Serbs and the Bosnian government to agree to a cease-fire and withdrawal of heavy weapons.

Simultaneously, at the grand strategic level, the U.S. and French governments began pressing for a NATO ultimatum to the Bosnian Serbs: "Cease fire and withdraw heavy weapons or we launch strategic air strikes against you." Although it supported his efforts to get Serb agreement, this was of serious concern to the UNPROFOR commander, who feared that NATO air strikes would make it impossible to continue his humanitarian mission and would endanger the lives of UN and non-governmental organization personnel. His reasoning followed this path: "If my mission is humanitarian, let me do it. If you want to change my mission to fighting the Serbs, I must withdraw UN personnel and abandon the humanitarian mission."

The commander proceeded urgently with his new mission in the hope of forestalling air strikes. He had already laid the groundwork. In previous discussions he had obtained the agreement of commanders of Serbian and government forces to four points: a cease-fire; withdrawal of Serbian heavy weapons to at least 20 kilometers from Sarajevo if not placed under UN control; the interposition of UN troops; and daily meetings to oversee implementation of these agreements. When the crisis broke, however, Bosnian President Alija Izetbegovic decided he would now make no agreement with the Serbs. Rather, his policy would be to take advantage of the outcry to persuade the international community to take action on the Bosnian government's behalf, specifically by lifting the arms embargo against it and taking military action such as air strikes against the Serbs.

The result was that the commander discovered that the president, who was at that moment being interviewed by CNN and declaring the need for air strikes against the Serbs, had told his commanders not to attend the UNPROFOR-arranged meeting.

Angrily the commander let it be known that if the Bosnian government did not attend the meeting, he would publicly blame Muslims for the market-square incident and for blocking a cease fire. Table 11 is our attempt at a card-table model of the commander's moment of truth at this juncture.

In table 11 the symbol ~ indicates that a character takes no position as to whether a particular card should be played or not; that is, its position is that the card may or may not be played. The Bosnian Serbs, for example, took the position (column **BS**) that UNPROFOR should not blame them for the market-square bomb nor call air strikes against them, while they themselves should not cease fire or withdraw weapons; in return they would not retaliate against UN personnel. They took no position as to whether UNPROFOR should blame the Muslims or the Bosnian government should cease fire; obviously they would have preferred both these cards to be played, but could hardly demand or expect them, given their own proposed actions. The numbers indicate preference rankings as before, with the most preferred future given the number 1, the next most preferred the number 2, and so on. In assigning preference rankings to columns containing entries with ~, the most probable decision is assumed. For example, it is assumed in this table that if the Bosnian Serb position, **BS**, were accepted, then the players' expectation in regard to the cards left unspecified ("Blame Muslims" and the Bosnian government's "Cease fire" card) would be that neither card would be played: UNPROFOR would not blame the Muslims and the Bosnian government would not cease fire. This is why the column **BS** is assigned the same preference ranking for each player as column *d*, even though it is compatible with either *t* or *d*.

	U	BS	BG	*t*	*d*
UNPROFOR	1	2	4	3	2
Blame Serbs	☐	☐	■	☐	☐
Call air strikes against Serbs	☐	☐	■	☐	☐
Blame Muslims	☐	~	☐	■	☐
BOSNIAN SERBS	3	2	4	1	2
Cease fire, withdraw weapons	■	☐	~	☐	☐
Retaliate against UN personnel	☐	☐	~	☐	☐
BOSNIAN GOVERNMENT	3	2	1	4	2
Cease fire	■	~	☐	☐	☐

LEGEND

■ means card is played
☐ means card is not played
~ means card may or may not be played
U is UNPROFOR position
BS is Bosnian Serb position
BG is Bosnian government position
t is threatened future
d is the default future
# indicates preference ranking (1 is most preferred)

Table 11. Bosnian government refuses to discuss a cease-fire.

The following assumptions are made in the model:

- The UNPROFOR position (column **U**) was that both sides should cease fire while U (UNPROFOR) remained impartial, not blaming or attacking anyone. Note that it was important for U that BG (the Bosnian government) as well as BS (the Bosnian Serbs) should cease fire. BG had more infantry, although fewer heavy weapons, than BS. Also, it often initiated incidents, even though it might lose militarily, to get world opinion on its side.

- The commander could assume that the BS position was as shown (column **BS**), even though BS had agreed to U's four points. This is because BS had agreed unwillingly, arguing that as BG forces outnumbered theirs, they needed to use artillery to respond to BG-initiated incidents. When BG rejected the four points, it could be assumed that BS would reject them also in favor of its previous position.

- The BG position was that they, the victims, should not be asked to cease fire. U should take their side, blaming BS and calling air strikes against them. BS could do as they wished, (i.e., BG took no position as to which BS cards should be played).

- The assumption behind the fourth column is that, by angrily interrupting the president's CNN interview and directly threatening him, the commander made the threatened future one in which he would publicly blame the Muslims and the bombardment would continue. This, of course, was very much what the Serbs wanted. The threatened future put no pressure on BS; however, it did pressure BG to accept U's position.

• The fifth column states that, if present policies and actions continued unchanged, there would be no cease fire and U would take no action.

Other Parties in the Background

There were at least two other important parties in the background of the drama we are analyzing: the Serbian government of Slobodan Milosevic and the Croatian government of Franjo Tudjman. At this point both were in transition from one policy to another. The Serbian government was attempting to get international sanctions against it lifted by distancing itself from the Bosnian Serbs, who had advanced so far with its backing. The Croats were engaged in U.S.-sponsored negotiations to restore their alliance with the Bosnian government against the Bosnian Serbs, having in the most recent phase of the conflict been taking land and expelling Muslims in unspoken alliance with the Serbs. Alliances were reforming against the Bosnian Serbs. This, we assume, increased their feelings of persecution and discrimination while making them more anxious to take Sarajevo while they could. At the same time, it meant that the Bosnian and Croatian governments played no active role in the confrontation we are analyzing.

Note that we are following and will continue to follow an important rule of realistic modeling, as stated in the play in chapter 6 by Mark Bright while advising the general. Mr. Bright points out that factors outside the model that influence things in it must be taken into account in making assumptions. They should not be ignored just because they are not explicitly named in the model.

Analysis of Table 11

What dilemmas face the characters in this model? Take the UN commander first.

U has no deterrence dilemma against BG. BG prefers column **U** to *t* (*t* would undermine international sympathy for BG); therefore U's threat, if credible, puts pressure on BG to accept column **U**. The result is that the UN commander's threat concerned Bosnian President Izetbegovic; he broke off his interview with CNN.

U does have a deterrence dilemma against BS, who greatly prefers *t* (with the West blaming the Muslims while their bombardment continues) to **U**. The result is that the Serbs were under no pressure, and would have been delighted if Izetbegovic had refused to attend the meeting.

U has a threat dilemma; it prefers not to blame the Muslims for the market-square bomb. That would mean giving up impartiality and possibly going against the grand strategic policy of NATO and the UN. The result is that the UN commander's anger, without which the Bosnian president might have dismissed his threat. This anger showed either or both of the following:

- That U's preferences were changing in favor of carrying out the threat (preference change)

- That U would carry it out regardless of whether or not it preferred to do so (irrationality).

U has an inducement dilemma in that *t* is arguably worse for U than **BS**, a fact that puts pressure on U to accept **BS**. The reaction was anger and defiance on the part of the UN commander as the BG position, with accompanying need to threaten *t*, puts pressure on him to abandon his objective and accept **BS**.

U has no positioning or cooperation dilemmas. It prefers its own position to the positions of the other characters and to any future it could reach from its own position.

U has a dilemma of trust in that both BS and BG would prefer unilaterally to defect from U's position by breaking the cease-fire. It is even preferable for both if both defect: preferable for BS because it wants to keep up the pressure on Sarajevo, preferable for BG because it wants to keep up pressure on the international community to intervene on its behalf. There is no immediate reaction to this dilemma. But note what happens next.

BG gives in to U's pressure and accepts U's position. BG and BS army commanders then attend a meeting with U. In this meeting all agree with U's position; however, even at the time U does not give much credence to this agreement. U does not decide to give up. Instead, U decides to monitor whether the agreement is observed while continuing, with positive emotion, to stress its desirability to both sides in an attempt to persuade them to keep it. All the signs are that U feels it has a trust dilemma.

Other Parties' Dilemmas

We have looked at U's dilemmas. What about other characters' dilemmas?

All the dilemmas facing the parties are set out in table 12. It is striking that the Serbs face no dilemmas. This is partly because their own position is what they want to see happen and is such that they do not need to trust anyone. They have no cooperation, trust, or positioning dilemmas. Also, the tug-of-war between U and BG has made the threatened future one they like, although the other two do not. They therefore have no inducement, threat, or deterrence dilemmas. They are sitting pretty.

	DETERRENCE	INDUCEMENT	THREAT	COOPERATION	TRUST	POSITIONING
UNPROFOR	√	√	√		√	
BOSNIAN SERBS						
BOSNIAN GOVERNMENT	√	√			√	

Table 12. Dilemmas facing characters in table 11.

The Bosnian government has three dilemmas:

- Deterrence: The threatened future, *t,* pressures no one to accept BG's position. Both U and BS prefer *t* to **BG**. The reaction was abandonment of position **BG**.

- Inducement: BG prefers **U** to *t*. The reaction was acceptance of position **U**.

- Trust: Even if U were to agree to BG's position, BG might doubt whether they would carry it out, given the UN commander's strongly expressed preference for remaining impartial and not taking sides. The reaction was possibly greater willingness to abandon position **BG**.

How UNPROFOR Got Angry—The Model That Preceded Table 11

It seems probable that the card, "Blame Muslims," in table 11 was thought of by the UNPROFOR commander as something he might actually do only when he realized the position the Bosnian government was taking. To see this, we draw up the card-table in table 13 showing the moment of truth before he thought up this card.

Both Bosnian sides here prefer t to U's position (BG does so because t places pressure on the international community to intervene on BG's side); therefore, U now faces a strong deterrence dilemma (i.e., it is pressuring no one to accept its position). U's reaction is anger and demonization of BG, motivating U to think up the card, "Blame Muslims," and make it credible as part of the threatened future t shown in table 11.

Note that if we look at the analysis formally, it might seem that U might equally well have demonized the Serbs. However, the Serbs had at least agreed to talk. The Muslim-dominated government was refusing to do so. Also, the Serbs already had been so demonized by the media and Western public opinion that U would need little rationalization, and hence little demonization, to justify turning against them.

	U	BS	BG	*t*	*d*
UNPROFOR	1	2	3	2	2
Blame Serbs	☐	☐	■	☐	☐
Call air strikes against Serbs	☐	☐	■	☐	☐
BOSNIAN SERBS	2	1	3	1	1
Cease fire, withdraw weapons	■	☐	~	☐	☐
Retaliate against UN personnel	☐	☐	~	☐	☐
BOSNIAN GOVERNMENT	3	2	1	2	2
Cease fire	■	~	☐	☐	☐

LEGEND
■ means card is played
☐ means card is not played
~ means card may or may not be played
U is UNPROFOR position
BS is Bosnian Serb position
BG is Bosnian government position
t is threatened future
d is the default future
# indicates preference ranking (1 is most preferred)

Table 13. What made UNPROFOR
angry? The situation preceding table 11.

This, then, is an example of emotion and rationalization leading to a change in the set of cards assigned to characters in a frame.

In table 14 we set out all the dilemmas in table 13.

• U has a deterrence dilemma (as said) and a trust dilemma (for the same reasons as in table 11).

• BS has no dilemmas.

	DETERRENCE	INDUCEMENT	THREAT	COOPERATION	TRUST	POSITIONING
UNPROFOR	√				√	
BOSNIAN SERBS						
BOSNIAN GOVERNMENT	√				√	

Table 14. The dilemmas facing characters in table 13.

- BG has a deterrence dilemma (no one is pressured to accept its position) and a trust dilemma (U would not want to implement BG's position, if it were agreed).

BG's dilemmas obviously gave it serious problems in this table; however, it had a strategy to overcome them, consisting of a strong appeal to the international community, which had it in its power to change the mission assigned to U and thereby change U's preferences. If it could get U on its side, this would be enough to achieve its position, as its position did not specify any particular actions on the part of the Serbs (i.e., any Serb actions would be compatible with its position).

AFTER THE BOSNIAN GOVERNMENT WAS BROUGHT INTO APPARENT COMPLIANCE

A commander's mission objective, expressed in terms of confrontation analysis, is to bring others into compliance with his position. Now at this point in our analysis, it might seem that BG, at least, has been brought into compliance.

A commander must, however, distinguish between apparent compliance (overt acceptance of his position) and actual compliance (in addition, the intention to implement that position). He must be aware that a character facing a cooperation dilemma may solve it by deceit (i.e., accepting a position while intending to act in non-compliance with it).

In this case the estimate is that BG has been brought into apparent compliance only, because if a cease-fire were implemented, they would prefer to break it. Although the UNPROFOR commander has achieved the bringing of BG into apparent compliance, BG now has a cooperation problem and the commander a trust problem in that BG will want to defect from position U.

Problems in Determining a Character's Position

Has BS also been brought into apparent compliance? A superficial reading might suggest so. We have said that all parties, including BS, have at this point formally accepted U.

This would be a mistake, illustrating the fact that determining a party's position by the rule, "It is what it says it is," may be harder in practice than it seems in theory. A real-life character has many subcharacters saying different, contradictory things. Even the same subcharacter may contradict itself. What do we do then?

Judgment is necessary, but a general rule in interpreting others' messages is, "Pay attention to their emotions." One thing we know is that a character whose emotional signals and rationalizations are predominantly negative toward others, angry, defiant, and so forth, must be taking a position opposed to theirs. We know this because such emotions are not appropriate to solving cooperation or trust dilemmas, which are the only ones left between parties that share a common position.

If a negatively emoting party formally states that it shares your position, it is contradicting itself by its negative rationalizations. If these are as open and public as its formal statement of agreement, if they come across more strongly than the formal statement, and if they state a clear alternative position, they should be read as stating the party's actual declared position.

In this light, consider the present case. At this point, the signals being sent by BS and BG are quite different. BG is showing signs (i.e., fear of the threatened future and a conciliatory attitude) of the impact of the pressures needed to make them comply. BS is showing opposite signs. The NATO ultimatum to the Serbs, "Cease fire and withdraw or we bomb you," was issued on the day that agreement apparently was reached. Instead of reacting with fear, depression, or conciliation, BS reacted with defiance, threatening to shoot down 40 NATO planes in the first wave of attack. More seriously, they threatened retaliation against UN personnel. This, together with their continuing strong arguments for column BS, meant that this was their position. They were not even in apparent compliance with U.

One interpretation, possibly that of the commander, is that the NATO ultimatum itself, by infuriating BS, had made them non-compliant. It is equally possible that BG's compliance made the BS non-compliance obvious. While BG was defying U, BS may have thought it had nothing to lose by seeming conciliatory.

In either case, Serbian defiance as now revealed means that the correct model of the moment of truth at this point is that in our next model, table 15. Here BS has de facto not agreed to U's position, even though formally it has done so. Only BG has agreed. In addition, the threatened future is that U will blame the Serbs and call air strikes against them. BS then will retaliate against UN personnel.

Modeling the Grand Strategic Drama

Note that the threatened future in table 15 comes from the NATO ultimatum rather than from the words of the UNPROFOR commander, who was strongly opposed to air strikes because he feared retaliation.

The NATO ultimatum emerged from a grand strategic drama going on in parallel to the operational drama represented in table 15. We attempt to model this higher-level drama in table 16, before discussing how it is linked to table 15. Again, our assumptions about the grand strategic model use published sources only, and readers who may know them to be wrong are asked to make allowances.

	U, BG	BS	*t*	*d*
UNPROFOR	1	2	3	2
Blame Serbs	☐	☐	■	☐
Call air strikes against Serbs	☐	☐	■	☐
Blame Muslims	☐	~	☐	☐
BOSNIAN SERBS	2	1	3	1
Cease fire, withdraw weapons	■	☐	☐	☐
Retaliate against UN personnel	☐	☐	■	☐
BOSNIAN GOVERNMENT	3	2	1	2
Cease fire	■	~	☐	☐

LEGEND
■ means card is played
☐ means card is not played
~ means card may or may not be played
U, BG is position of UNPROFOR and Bosnian government
BS is Bosnian Serb position
t is threatened future
d is the default future
# indicates preference ranking (1 is most preferred)

Table 15. Ultimatum to the Bosnian Serbs.

First, what cards are in play and who holds them?

The United States controls the card, "Bomb Serbs into compliance." Such bombing would be done by NATO, but this card is assigned to the United States because, first, the United States has the necessary military assets and, second, NATO as a whole, apart from Britain, is prepared to support the United States in this.

Russia has the important card, "Back Serbs against NATO." Russia's pro-Slav sympathies and desire to have a friendly Slavic nation controlling former Yugoslavia incline it to play this card, if necessary, to offset what it sees as Western anti-Serb bias.

The Bosnian Serbs have the same cards in this confrontation as at the operational level (see table 15).

Finally, Britain has the card, "Publicly oppose bombing," a card it is inclined to play to keep a reputation for impartiality and lessen the danger to its vulnerable troops engaged in a humanitarian mission.

What positions are parties taking?

The United States and Britain (B) take somewhat differing positions, but because their positions are close, we have collapsed them into one. We have done this by leaving it open whether, under the joint position, B would publicly oppose bombing. B proposes to do so (to show impartiality and protect its troops), and the United States is willing to let it do so. The United States, Britain, and the rest of NATO agree that the Bosnian Serbs should cease fire and withdraw weapons, obviating the need for the United States to bomb the Serbs or for the Bosnian Serbs to retaliate against UN personnel.

Where Britain differs from the United States and the rest of NATO is in being opposed to the U.S. fallback position, "Bomb Serbs into compliance." Britain's own fallback position is to continue publicly to oppose bombing. Britain takes this fallback position out of preference (to show impartiality and protect its troops) and to deter the United States from taking its fallback position by making that position less comfortable for the United States (and so helping to give the United States inducement and threat dilemmas that might lead it to change its fallback position). Britain is not taking this fallback position as a way of inducing the Bosnian Serbs to accept the joint position U,B.

Russia (R) at this stage has not taken a clear position on any issue except that it is against NATO bombing of BS (it has agreed to close air support, but excluded punitive air strikes). It has not taken the position of backing the Serbs against NATO, but is implicitly threatening to do so if NATO bombing is implemented.

Meanwhile, the United States, Britain, and the rest of NATO take no position on whether Russia should back the Serbs against NATO. That, they consider, is up to Russia. But if Russia should carry out its implicit threat to back the Serbs while NATO is bombing them, there would be serious concern about a deterioration of relations between Russia and the United States, even fear of a new East-West confrontation arising over the Balkans.

The threatened future, *t*, is that the United States commences to bomb the Serbs into compliance (the action would be taken by NATO, but our assumption is that the decision would be taken by the United States); Russia responds by backing the Serbs against NATO, with all the geopolitical risks that entails; the Bosnian Serbs respond by retaliating against UN personnel; and the British publicly oppose bombing. Whether or not it is credible, this is the future implied by the parties' current fallback positions.

General Positions

In table 16 we have made free use of the symbol ~ to represent general positions (i.e., to show that a party has not taken a clear position as to whether certain cards should be played). The symbol ~ means that a card may or may not be played. Hence, when incorporated into a character's position, it means that the character takes no position as to the playing of that card.

We must use this symbol here in order to be realistic. Your position is something you are willing to go to the threatened future for. If you are not willing to go that far, then at the moment of truth it becomes clear that effectively you are not taking a position on that card (i.e., your position is compatible both with it being played and with it not being played). This is the case with the U.S. attitude to Britain publicly opposing bombing; the United States does not like it, but it is known it will accept it. Similarly with the Bosnian Serb attitude to Russia backing it against NATO; the Bosnian Serbs would like it, but are not demanding it (at least not in this model, although if we modeled Serb-Russian negotiations, we might find the Bosnian Serbs demanding it). Russia has not taken a firm position on any issue except NATO bombing of the Serbs.

Use of the symbol ~ means that the different positions of different parties may be compatible (e.g., in table 16 Russia's position is compatible with both of the other two, although they are incompatible with each other).

	U,B	BS	R	*t*	*d*
U.S.	1	3	3	2	3
Bomb Serbs into compliance	☐	☐	☐	■	☐
RUSSIA	1	2	2	3	2
Back Serbs against NATO	~	~	~	■	☐
BOSNIAN SERBS	3	1	1	2	1
Cease fire, withdraw weapons	■	☐	~	☐	☐
Retaliate against UN personnel	☐	☐	~	■	☐
BRITISH GOVERNMENT	4	2	2	3	2
Publicly oppose bombing	~	~	~	■	■

LEGEND

■ means card is played
☐ means card is not played
~ means card may or may not be played
U, B is U.S. and British position
BS is Bosnian Serb position
R is Russian position
t is threatened future
d is the default future
# indicates preference ranking (1 is most preferred)

Table 16. Grand strategic pressure on the Bosnian Serbs.

Linkages Between Models—Use of Context Cards

We now have models of two confrontations, at the operational and grand strategic levels, that are clearly linked.

First, we can see that the reason the Bosnian Serbs have the same cards in each model is that they are playing simultaneously in both confrontations. At the grand strategic level they are sending messages to the United States, Russia, and Britain, while at the operational level they are negotiating with the UNPROFOR commander's staff and through it, indirectly with the Bosnian government.

Second, it is clear that the U.S. fallback position in the grand strategic game ("Bomb the Serbs into compliance") partly dictates the UNPROFOR commander's fallback position at the operational level (i.e., it means his fallback must contain the card, "Call air strikes against Serbs"); however, it does not dictate which other cards he must play. Because he is the man on the spot, responsible for interacting with the Bosnian Serbs and having immediate responsibility for deciding if they have complied, he must decide at what exact point this fallback position should be implemented. Realistically, we should regard him as holding and negotiating with the card, "Call air strikes," while knowing (and knowing that the Serbs know, etc.) both that he has been directed to threaten its use and that his decision to play it or not may be overridden by a U.S. decision in the grand strategic game. Thus he plays with this card knowing that at any time he may be tapped on the shoulder and told what to do with it, but fearing that if he cannot get agreement, he must use it.

The Bosnian Serbs' tactic makes use of this. They plan to negotiate with him while simultaneously sending messages (both through him and through the Russians) in the grand strategic game. Their aim is to make the threatened future in that game so bad for the United States that it will back off from its fallback position, thereby making UNPROFOR back off from its fallback position in the operational game. The Bosnian Serbs plan to do this partly by threatening to retaliate against UN personnel, partly by getting the Russians to threaten to back them against NATO, and, to a lesser extent, partly by the threat of British opposition to bombing.

The UNPROFOR commander's tactic, on the other hand, is driven by the fact that while opposed to bombing, both personally and in line with his national government's policy, to avert it he must credibly threaten to do it. While credibly threatening, he can at the same time use his discretionary powers to try to water down the ultimatum (e.g., let the Serbs keep their weapons in place provided he can monitor them), thereby making conciliatory adjustments to his position to make it more acceptable to the Serbs. He desperately tries both tactics to get an agreed position before the NATO ultimatum expires; however, his bargaining position is strengthened by his ability to use the "nice-guy-mean-guy" argument that, even though he does not like bombing, he must order it if the Serbs do not comply.

Thus the models are linked first, because the Bosnian Serbs are playing in both games at once, using the same cards, and secondly, the UNPROFOR commander's choices in one game are partly dictated by those made in the other, although he does exercise some initiative as a subordinate; that is, he has cards he can play or not play and can choose his negotiating strategy.

What does the linkage do to characters' preferences? For example, to know if the United States and Britain have a deterrence dilemma, we need to decide if the Bosnian Serbs prefer *t* to **U,B** in table 16. Is this the same as whether they prefer *t* to **U, BG** in table 15?

It is the same question, and will receive the same answer, provided we follow our constant advice, which is to consider relevant factors outside the model when making assumptions within a model.

In answering the question relative to table 16, we need to recall that in table 15 UNPROFOR is threatening to play the card, "Blame the Serbs," as part of *t* while at the same time seeking to make compliance with **U, BG** less onerous for the Serbs.

In answering the same question in the context of table 15, we need to recall that in table 16 the Russians are indicating they will play the card, "Back Serbs against NATO," as part of *t*, and the British are indicating they will publicly oppose bombing.

This consideration of external factors can be done formally by adding context cards to a model. This means adding cards below a line labeled "Context," as in the detailed model of table 7. We can add cards showing the assumptions made about any external factors. To show how this works, table 17 is the same as table 15, but with added contextual assumptions about what would be happening in table 16. These assumptions are those stated above.

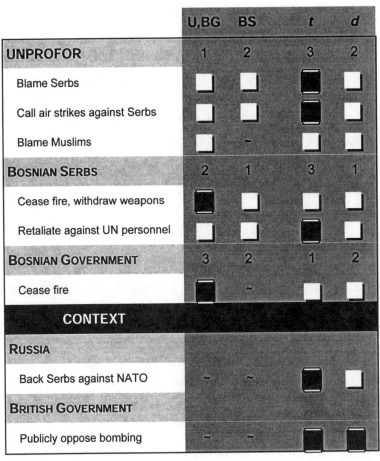

LEGEND
■ means card is played
□ means card is not played
~ means card may or may not be played
U,BG is position of UNPROFOR and Bosnian government
BS is Bosnian Serb position
t is threatened future
d is the default future
# indicates preference ranking (1 is most preferred)

Table 17. Ultimatum to the Bosnian Serbs with
context cards added from grand strategic model.

Dilemmas in the Grand Strategic and Operational Dramas

We have looked at players' positions and tactics in the grand strategic and operational confrontations as NATO presents its ultimatum to the Bosnian Serbs. We have not spelled out the dilemmas they face.

The dilemmas in the grand strategic confrontation are shown in table 18.

The United States and Britain face an inducement dilemma and the United States faces a threat dilemma in that *t* could lead to an East–West confrontation, as well as retaliation against UN personnel; therefore, it seems that the United States and Britain would prefer the Bosnian Serb position to *t* (this is their inducement dilemma) and the United States would prefer not to bomb the Serbs (its threat dilemma). Britain's reactions include fear of an East–West confrontation (and of endangering British troops on the ground); this fear was conveyed to the Bosnian Serbs by David Owen, who was moved by it to conduct diplomacy against air strikes (Silber and Little, 1996, pp. 311-313; Owen, 1996, chap. 7). The U.S. reaction included rationalization of preference for *t* over column **BS** by issuing an ultimatum and following the ultimatum, arguing that NATO credibility is now at stake.

The United States and Britain face a deterrence dilemma in relation to the Bosnian Serbs, who see an East–West confrontation in which Russia takes their side (this is the BS view of *t*) as likely, in their view, to end with the West making compromises in their favor, and hence as being preferable to position **U,B**. U.S. and British reactions include attempts to get Russia

	DETERRENCE	INDUCEMENT	THREAT	COOPERATION	TRUST	POSITIONING
U.S.	√	√	√		√	
Russia			√			
Bosnian Serbs						
British Government	√	√			√	

Table 18. Dilemmas in table 16.

to threaten the Bosnian Serbs with lack of support. U.S. President Clinton phoned Russian President Yeltsin and British Prime Minister Major visited Yeltsin in Moscow to request Russia to pressure the Bosnian Serbs to accept the position **U,B**.

The United States and Britain face a trust dilemma because they cannot trust the Bosnian Serbs to continue to implement the agreement **U,B**, even if they should seem to comply (i.e., even if they begin to implement it in the short term).

Russia faces a threat dilemma in that Russia, like the United States and Britain, also fears a renewed East–West confrontation and would prefer not to play the card, "Back Serbs against NATO," as part of *t*. The reaction included conflicting conciliatory and angry signals from Moscow, ending, after a visit by British Prime Minister to the Russian president, with acceptance by Russia of a joint Anglo-Russian position compatible with the **U,B** position in table 15 and a threat to the Bosnian Serbs that if they rejected this, Russia would not back them against NATO.

The next step in the resolution process is that Russia, in an attempt to eliminate its threat dilemma, shifts its position and brings about a new moment of truth; however, before looking at this, we will look at the dilemmas found in table 15, where the UNPROFOR commander negotiates at the operational level with the Bosnian Serbs. Table 17 is the same as table 15 except that context cards have been added to show influences from the grand strategic level.

The dilemmas found at this level are shown in table 19.

	DETERRENCE	INDUCEMENT	THREAT	COOPERATION	TRUST	POSITIONING
UNPROFOR	√	√	√		√	
Bosnian Serbs						
Bosnian Government	√			√	√	√

Table 19. Dilemmas in tables 15 and 17.

U (UNPROFOR) and BG (the Bosnian government, now sharing U's position) face deterrence and trust dilemmas. Column *t* puts BS under no pressure to accept column **U,BG** (this is the deterrence dilemma facing U and BG.) In any case, because BS already have formally accepted it, any further agreement on their part to do so could hardly be trusted (the trust dilemma facing U and BG). Reactions include U attempts to improve **U,BG** from the BS viewpoint by reinterpreting the agreement to allow BS guns to stay in their positions subject to monitoring. For BG, on the other hand, these dilemmas are welcome as arguments in favor of abandoning the position **U,BG** and resorting to a position it much prefers (i.e., the threatened future *t*); therefore, BG urges strict compliance with the agreement **U,BG,** hoping for non-compliance and the implementation of *t*.

U also has inducement and threat dilemmas. The UNPROFOR commander himself openly states that he prefers not to call air strikes against the Serbs, and his dislike of this is supported by the grand strategic fear of East-West conflict. The reaction includes a mixture of anger and despair on the part of U, which nevertheless keeps trying to get an agreement.

BG also faces a cooperation dilemma (it prefers not to keep to the agreement) and a positioning dilemma (it prefers **BS** to **U,BG**) because under **BS** it would hope for eventual Western intervention. The reaction is that these dilemmas are quite welcome for BG. It has been forced into unwilling acceptance of column **U,BG**. If this column is implemented, BG no doubt intends to defect from it in due course, breaking the cease-fire, perhaps gaining some ground, but provoking further BS attacks on the basis of which it can make further appeals to international opinion.

How Real Pressure Was Brought to Bear on the Serbs

Note that in these models the Bosnian Serbs face no dilemmas. Their position **BS** is preferred by them and stable for others, so that they have no cooperation, trust, or positioning dilemmas. Also, *t* is attractive to them because it would involve Russian support, so that they have no inducement or threat dilemmas. Also, they have no deterrence dilemmas because *t* is feared by Russia and the Western players, while BG likes **BS**.

All this changes after the British Prime Minister's visit to Moscow and Russia's subsequent agreement to adopt a joint position with Britain and put pressure on the Serbs to accept it. In table 20, the British-Russian joint position is shown in column **B,R**. Here Russia proposes to give its backing to the Serbs in return for Serb agreement to cease fire and withdraw weapons, while Britain publicly opposes bombing. Note that this is compatible with the U.S. position. This is not to say that the United States likes all its elements; it dislikes Russian backing for the Serbs and British opposition to bombing. Its compatibility with the U.S. position merely means that the United States is prepared to accept it, rather than go to its fallback position.

U, B, and R are now united against BS in demanding that the Serbs cease fire and withdraw; however, the significant point is that Russia's fallback position has shifted. If BS rejects **B,R**, then Russia no longer will back the Serbs. This becomes clear to the Bosnian Serbs when Russian Ambassador Churkin visits the Serbs and suggests position **B,R**. On the same visit, Russia assures the Bosnian Serbs of its willingness to back the Serbs against NATO if the Serbs will agree to position **B,R** by undertaking to send Russian troops to supervise the withdrawal of Bosnian Serbs heavy weapons. This changes everything for BS. It is now implicit that, if they refuse this offer, Russia will not back them against NATO. The threatened future *t* is now one in which the world, including Russia, is against

	U	BS	B,R	t	d
U.S.	1	3	1	2	3
Bomb Serbs into compliance	☐	☐	☐	■	☐
RUSSIA	1	2	1	3	2
Back Serbs against NATO	~	~	■	☐	☐
BOSNIAN SERBS	2	1	2	3	1
Cease fire, withdraw weapons	■	☐	■	☐	☐
Retaliate against UN personnel	☐	☐	☐	■	☐
BRITISH GOVERNMENT	1	2	1	3	2
Publicly oppose bombing	~	~	■	■	■

LEGEND
■ means card is played
☐ means card is not played
~ means card may or may not be played
U is U.S. position
BS is Bosnian Serb position
B,R is the British and Russian position
t is threatened future
d is the default future
# indicates preference ranking (1 is most preferred)

Table 20. Real pressure brought to
bear at last on the Bosnian Serbs.

them. Also, *t* now holds fewer fears for the United States and Britain (no more fear of a new East–West confrontation). Britain's fear of Bosnian Serb retaliation against UN personnel remains, but affords little comfort because U.S. troops are not threatened, and it is the United States that controls the card, "Bomb Serbs."

The dilemmas arising in table 20 are shown in table 21. The United States and Russia have only the dilemma that they cannot trust the Bosnian Serbs to keep a cease-fire for long, after they have agreed to one. Britain has, in addition, the inducement dilemma that it prefers the Bosnian Serbs' position to t (because of the threat t poses to British troops). This is a problem for Britain, but does not help the Bosnian Serbs because it is not Britain's fallback position, but that of the United States, which makes t unpleasant for the Bosnian Serbs. The Bosnian Serbs have a deterrence and an inducement dilemma of a kind that they can see no way to solve.

The Bosnian Serb's response is to seize upon the offer by B,R as a position it even prefers to its own because it includes Russian backing. BS is joyful; arriving Russian troops are greeted by cheering crowds.

	DETERRENCE	INDUCEMENT	THREAT	COOPERATION	TRUST	POSITIONING
U.S.					√	
Russia					√	
Bosnian Serbs	√	√				
British Government		√			√	

Table 21. Dilemmas in table 20.

There is now agreement by all parties on the position B,R (the intersection of the two compatible positions U and B,R). This is reflected in the operational drama, as shown in table 22, where all parties now take the same position. The remaining dilemmas are those of cooperation and trust, indicated by the question marks on the cease fire cards played by the Bosnian Serbs and the Bosnian government. Note that the Bosnian government here is accepting the position column that is worst for it among the three columns shown. This arose, as we have seen, because the Bosnian government was threatened with a still worse column (not shown here) where UNPROFOR would blame it for the market square massacre; however, the fact that it prefers either of the other two columns shown to the one it is accepting emphasizes its cooperation dilemma (and UNPROFOR's trust dilemma). It cannot be trusted to stick to the agreement. Reactions to these dilemmas include the following: deceit on the part of the Bosnian Serbs and the Bosnian government, hiding their intention to break the cease-fire; disbelief on the part of U, leading to the UNPROFOR commander monitoring the cease-fire and trying to find sanctions that will make defection unattractive and benefits that will make the agreement attractive to the Bosnian Serbs and the Bosnian government.

	U,BG, BS	*t*	*d*
UNPROFOR	1	3	2
Blame Serbs	☐	■	☐
Call air strikes against Serbs	☐	■	☐
Blame Muslims	☐	☐	☐
BOSNIAN SERBS	2	3	1
Cease fire, withdraw weapons	?	☐	☐
Retaliate against UN personnel	☐	■	☐
BOSNIAN GOVERNMENT	3	1	2
Cease fire	?	☐	☐
CONTEXT			
RUSSIA			
Back Serbs against NATO	■	☐	☐
BRITISH GOVERNMENT			
Publicly oppose bombing	■	☐	☐

LEGEND
■ means card is played
☐ means card is not played
U,BG,BS is position of all three parties
t is threatened future
d is the default future
? means playing this card is not preferred
# indicates preference ranking (1 is most preferred)

Table 22. Resolution (deceptive) of the
operational drama, with grand-strategic context.

What This Analysis Has Achieved

In this chapter we analyzed the dilemmas faced by the UNPROFOR commander and certain characters he interacted with in 1994. We tried to show that their reactions to these dilemmas can explain the emotions felt and initiatives undertaken by the characters.

We explained how and why the UNPROFOR commander, faced with Bosnian government intransigence, reacted angrily, thereby making credible a newly minted threat to place public blame on the Muslims. This brought the Bosnian government into fearful, apparent compliance with UNPROFOR's position. There remained the problem that the Serbs, although they had declared themselves willing to comply, now showed unequivocally by their actions, public statements, and defiant emotional attitude that their position now was one of non-compliance. They were eventually brought into apparent compliance by UNPROFOR's threat of bombardment; however, this threat was made credible only by grand strategic action (the NATO ultimatum) and became adequate only when Russia, again acting on the grand strategic level, made it clear that the Serbs would not have its backing against NATO. After it was made credible and adequate, the threat achieved apparent Serb compliance, although the Serbs remained defiant. The Serbs took consolation from the fact that Russia, thought to be their ally, had become involved. The eventual attitudes of both Serbs and Muslims indicated that their compliance was more apparent than real.

Our detailed explanations of these events are post hoc; moreover, explanation by itself is of little value.

The important question is whether this analysis, if done at the time, would have helped the UNPROFOR commander achieve his mission objectives. This question is addressed in the next chapter.

SUMMARY OF CHAPTER 8

To illustrate how the approach throws light on a real-world confrontation, this chapter reports an analysis carried out for DERA of the episode in February 1994 when the UNPROFOR commander was tasked with stopping the Bosnian-Serb bombardment of Sarajevo. The analysis was drawn from published sources only. Interviews were conducted with officers who had served at various levels in the Bosnian mission, but not at this stage or during this particular crisis. An explanation of the behavior and reactions of the various parties is given using drama-theoretic principles.

CHAPTER 9

WOULD THIS ANALYSIS HAVE HELPED?

Analyzing a confrontation, as in the previous chapter's analysis of the Sarajevo crisis of February 1994, does not answer the question, "How might confrontation analysis have helped the commander's operation-level decisions, if the method had been available to him at the time?"

To answer this, we will need to assume that our analysis is broadly correct, despite the fact that it is based on published accounts written after the event, rather than on participants' on-the-spot judgments. Not only are published accounts a poor guide to how things seemed at the time, they also have the disadvantage of concentrating on events at the operational and grand strategic level, rather than the tactical level. The effect is that we will not be able to say much about how the commander might have devolved his operational strategy to lower levels.

Despite these difficulties, we will project the analysis backward, beginning with the situation after the market square bomb, and imagine what might have happened if the commander had been able to formulate and follow a confrontation strategy guided by our analysis.

Rerunning the History of a Crisis

On February 5, 1994, let the commander begin by building the model in table 13. At this point, he has realized that the Bosnian president will not allow his officials to attend cease-fire talks because he is trying to use the situation to bring about international intervention on his side.

Determining End States

In the course of building this model, the commander has defined his end state. This is column **U**, the future in which BS (Bosnian Serbs) ceases fire and withdraws weapons, BG (Bosnian government) also ceases fire, U (UNPROFOR) remains impartial (not blaming Serbs nor calling air strikes), and there is no retaliation against UN personnel.

This is the same as U's position, although a character's end state is not always its position. Your position is your overtly declared end state, what you propose to the other characters as a solution. We have seen that a character with a cooperation dilemma has a temptation (perhaps working with others) to move from its position, if accepted, to another future that it prefers. It may solve this dilemma by deceit, so that its secret intention and actual end state is to carry out its temptation, not its position. Again, a character may have reluctantly accepted a position while planning for it to be rejected in favor of another. In this case, the latter rather than the former will be its actual end state.

In table 13, it appears that each party's end state is identifiable with its position; thus, the commander, in building this model, has determined not only his own end state, but the other parties' too. He has identified their goals.

End states found in a model of a common reference frame, which is what a card-table is, are necessarily simple. Simplicity is desirable. It is also realistic. In a confrontation, parties' end states, representing what they want others to accept, are necessarily simple. At the same time, the commander's staff will look at these end states in more detail to explore their implications and consequences. They do this by adding extra cards, both internal (i.e., above the context line) and external (below it), as in table 7. Simplicity, clarity, and certainty are lost in this process, but important points may be uncovered and difficult questions raised. Going into detail in this way is also necessary to devolve a strategy to the tactical level.

To find end states for a mission in a card-table model, you obviously must model that mission, and the mission must be a confrontation (i.e., an Operation Other than War [OOTW]). In our example, the commander has modeled the mission he has just received, not his previous and continuing humanitarian mission, and has determined his end state for this new, obviously confrontational mission. Lower-level, devolved models will enable determination of end states for lower-level missions.

Conversely, to determine his superior's end state, and the end state of his superior's superior, the commander models the grand strategic drama by building the model in table 16. In doing so, he identifies his superior's intent by modeling how Britain (his country), the United States (supported in this by the rest of NATO), and Russia see the application of pressure to the Bosnian Serbs at the grand strategic level. Observe that his superior's intent appears, not as a clearly defined mission, but as a position in a political drama being played out between national governments. The end state of this conglomerate superior is column **U,B** in table 16.

The commander's end state is, in general, a more detailed implementation of his superiors' end state, just as the end states in strategies devolved from his strategy will be more detailed implementations of his own end state. In moving directly to achieve his own end state he is helping his superiors to achieve theirs and using his subordinates to implement matters of detail within his.

In the case we are considering, the commander finds a problem with his superiors' model of the situation. They are implicitly assuming that the Bosnian government shares the position of the U.S. and British governments, or at least accepts what these governments think is good for it. BG (the Bosnian government) is not even a player in the grand strategic model. This reflects the way the governments see this moment of truth. They see themselves as negotiating with each other over how to bring the Bosnian Serbs into line. BG does not enter into it.

How should the commander deal with the fact that, contrary to his superiors' assumptions, BG is planning to take advantage of their blindness by refusing to settle, knowing that the Serbs will be blamed for the resultant perpetuation of the bombardment?

In light of his superiors' intent, the answer is clear. He must use his own initiative to bring BG into line.

Centers of Gravity and Sequences of Decisive Points

Use of some familiar military terminology may help describe what the commander does. We can say he attains his end state by finding a sequence of decisive points by which to destroy the centers of gravity of non-compliant parties.

The significance of looking for other parties' centers of gravity is this. Chapter 7 contained a description of the commander as cycling through a number of dilemma eliminations to achieve his end state; however, formal analysis of dilemmas is never enough. It is essential to look at the real-world context in which dilemmas arise to see how to eliminate them in the desired manner (i.e., without pressuring others into further escalation or into adoption of yet more contrary positions). This can be described as a matter of defining and attacking the center of gravity of their non-compliance.

As an example, the center of gravity of the Bosnian government's non-compliance could be defined as its project of working on Western public opinion to bring the West to its side. This center of gravity was effectively attacked by the UNPROFOR commander's threat to blame the Muslims for the market-square bomb.

This illustrates the point that to define a non-compliant party's center of gravity, it is helpful to first decide what its policy is. Is its position non-compliant? If so, it must have a policy by which it hopes to get acceptance of that position, and that policy must be overthrown, as the UNPROFOR commander overthrew the Bosnian government policy of prolonging the Serb bombardment to arouse international opinion against the Serbs. On the other hand, it may be that its overt position is compliant, and its non-compliance lies in its presumed intention to defect from it. In that case, it is the policy of defection that must be overthrown. Its center of gravity may be the belief that defection will be possible or the belief that it will be profitable.

In seeking to define centers of gravity, it usually is necessary to analyze (formally or informally) the internal drama going on within other players. The present policy of non-compliance has been arrived at as the resolution of a confrontation between various internal subcharacters. Our aim must be to launch an attack on certain agreed underpinnings of that policy, thereby throwing it into crisis and bringing about a shift to a policy of compliance. The commander's task is to eliminate dilemmas at his own level, working where necessary through subordinates, coordinating with others on his own level, and requesting assistance as necessary from higher levels.

Each round of dilemma elimination at his own level may be described as a decisive point for the commander. His confrontation strategy as a whole may be described as a plan to move through a sequence of decisive points, success at all of which will eliminate non-compliant centers of gravity and bring all parties into sincere compliance, so achieving his end state.

Planning and Following a Confrontation Strategy

To show how he proceeds, we will describe several cycles of dilemma elimination. These will illustrate both how he might have planned a confrontation strategy and how he might have carried it out.

Step 1: We have said that he starts by building table 13. The dilemmas found here are set out in table 14. BS has no dilemmas. Both U and BG have deterrence dilemmas and trust dilemmas. Neither BS nor BG is under pressure to accept the commander's position.

Step 2: The commander must solve his deterrence dilemma against at least one of the other parties (i.e., bring pressure on at least one). Now he knows that his superiors are thinking how to bring BS (the Bosnian Serbs) into line. They are not even thinking about BG; therefore, he decides that he must deal with BG.

Now BG's deterrence dilemma means that its position is weak at the operational level because it has no means of inducing the commander or BS to accept its positive position (column BG). Its strategy is to remedy this by operating on the grand strategic level to change the commander's preferences from above. He can forestall this if he can overcome his deterrence dilemma toward BG using a credible, useful threat that will induce it to comply. (He may not be able to bring it into true compliance because it may still intend to defect from his position, but apparent compliance is a first step.)

The commander brainstorms with his staff for possible cards. The commander considers the BG center of gravity is its desire to swing international opinion on its side; therefore, an appropriate card to pressure BG with is, "Blame Muslims for market square bomb," a card suggested by the evidence that they were responsible. He provisionally adds this card to his hand.

Having done so, he builds table 11, which includes this new card in his fallback position. He finds, on analyzing this table, that he has solved his deterrence dilemma but given himself inducement and threat dilemmas (in that he prefers not to use the new card). He decides to overcome them either by showing anger (as was actually done in 1994) or in some other way found by brainstorming and role-playing the problem (e.g., he might represent Muslim blameworthiness as a logical deduction from BG's unwillingness to adopt the position U).

Solving his own dilemmas will, he decides, heighten BG's inducement dilemma to the point where it can solve it only by accepting his position.

Step 3: At this step he successfully carries out, or, in planning mode, rehearses, these dilemma-eliminations. Rehearsal methods described in chapter 10 are a good way of testing this phase of a confrontation strategy.

Step 4 consists of a return to Step 1. Because the planned dilemma-elimination has succeeded (or, in planning mode, has been assumed to have succeeded), in shifting BG's position to overt, if unreliable, acceptance of his position, the commander builds the model in table 17.

This model has context added from the grand strategic drama because this now seems to be of overriding importance. To establish this context, the commander also builds table 16, thereby modeling the problem faced by his superior.

In the new model at his own operational level, he faces up to the problem of bringing BS into apparent compliance. He identifies the dilemmas in the model as being those in table 19. He finds that he faces dilemmas of deterrence, inducement, and threat, as well as trust. BS has no dilemmas. BG has dilemmas, but because BG is still clearly aiming for a position other than the one it is overtly taking, it welcomes them, regarding them as arguments against its position and in favor of the position it would rather be taking.

Step 2: What can the commander do? In neither the operational nor the grand-strategic level table does BS now face any dilemmas. No pressure is being brought on BS.

This is disconcerting. The commander decides to use brainstorming and role-playing to search for the BS center of gravity. He finds that BS does not fear t in either model because BS sees R (Russia) backing it if NATO attacks. It seems, then, that BS's belief in Russian backing is the BS center of gravity. From his superior's model in table 16, the commander finds that R has an inducement dilemma and a threat dilemma in backing BS in the event of NATO air attacks.

That is useful information. The commander decides to use it to attack the BS center of gravity (after checking with his superiors, which he needs to do because the tactic of using these Russian dilemmas has grand strategic implications). His plan is to point out to BS that R will certainly not back it when the chips are down. Getting this message across effectively should eliminate his (UNPROFOR's) deterrence dilemma and give BS an inducement dilemma by making BS prefer his position to t.

Simultaneously, he plans to eliminate his own inducement and threat dilemmas and give BS a deterrence dilemma by pointing out that NATO is committed by its own ultimatum, and cannot now back down. NATO, unlike the UN, cannot afford to lose credibility. NATO's commitment is such that even the threat of retaliation against UN personnel is not sufficient to make column BS preferred to *t*; such a threat merely angers Western public opinion and increases the U.S. preference for *t* over the Bosnian Serbs' position, and, he will point out, U.S. preference, not British, is what makes *t* credible. He plans to point out that Russia knows all this, and therefore will itself back down; this is why it has not even taken a firm position. He will point out the inevitable conclusion that Bosnian Serbs' intransigence will lead to NATO air action plus withdrawal of Russian support. He will try to get this message reinforced by all sources in contact with the Bosnian Serbs, thus achieving unity of effort. For good measure, to take away any crumb of comfort the Bosnian Serbs may find in the threatened future *t*, he will add the assurance that while Britain does at present publicly oppose bombing, it will not do so if NATO takes action.

The commander builds these plans from analysis of dilemmas in conjunction with appreciation of the attitudes of the Bosnian Serbs. He has one problem, though. To overcome his inducement dilemma, he must find some way of dealing with his own reluctance to use air strikes.

He does not like air strikes. He fears their damaging effect on his ongoing humanitarian mission and prefers not to risk the lives of UN personnel. Yet his analysis makes it clear that to get rid of his inducement dilemma and give the Bosnian Serbs a decisive deterrence dilemma he must find some way of preferring *t* to the Bosnian Serbs' position. He finds an answer to this dilemma in the principle that he must try to implement his superior's intent. His superior, in this case, is a congeries of governments swayed by media-hyped public opinion and arriving at decisions through a process of political confrontation. Right or wrong, that is his superior. It is not, in this case, manifestly wrong. It is clear that the firm line being taken by the United States and supported by the rest of NATO (apart from his own government) means that his superior is swinging in the direction of maximum pressure to make the Bosnian Serbs comply; therefore, clearly his effective preference, mandated by his superior's intent, is for *t* rather than **BS**, much as he dislikes *t*.

BS simply must give in.

Step 3: BS does give in (or its role-players do, if we are still in planning mode). Another decisive point is passed. This occurs because U convinces BS that the true model it faces if it does not shift its position is that in table 23. Here, *t* involves neither R backing the Serbs nor B publicly opposing bombing, but continuing air strikes preferred by U to **BS**.

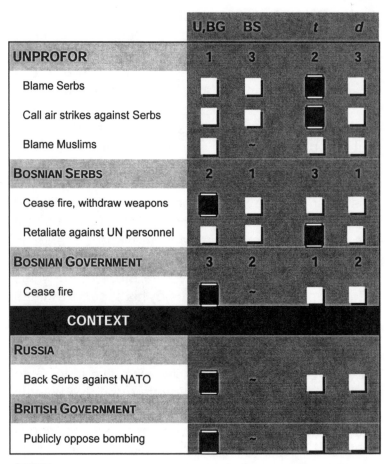

	U,BG	BS	t	d
UNPROFOR	1	3	2	3
Blame Serbs	☐	☐	■	☐
Call air strikes against Serbs	☐	☐	■	☐
Blame Muslims	☐	~	☐	☐
BOSNIAN SERBS	2	1	3	1
Cease fire, withdraw weapons	■	☐	☐	☐
Retaliate against UN personnel	☐	☐	■	☐
BOSNIAN GOVERNMENT	3	2	1	2
Cease fire	■	~	☐	☐
CONTEXT				
RUSSIA				
Back Serbs against NATO	■	~	☐	☐
BRITISH GOVERNMENT				
Publicly oppose bombing	■	~	☐	☐

LEGEND

■ means card is played
☐ means card is not played
~ means card may or may not be played
U, BG is position of UNPROFOR and Bosnian government
BS is Bosnian Serb position
t is threatened future
d is the default future
# indicates preference ranking (1 is most preferred)

Table 23. Ultimatum to the Bosnian Serbs
with threatening grand strategic context.

THE RERUN COMPARED TO THE REALITY

Observe that this rerun of the past achieves the same end result as the reality, but achieves it in a different way.

In reality, U got BG to overtly accept U's position by threatening to blame the Muslims for the market square bomb. This achieved the first round of dilemma elimination described in our rerun; however, rather than being done in a planned and thoughtful way, with an adequate appreciation of risks and ramifications, the elimination was done impulsively and emotionally. The commander got angry and made his threat credible in that way.

The commander's instincts were right, and he was successful; moreover, when there is little time to act, action must be fast and decisive. As a whole, U (including the commander, his staff, and UNPROFOR coevals) would have benefited from a clearer, more communicable understanding of what he was doing to win a confrontational battle. A major benefit might have been to win in such a way as to enlist more willing support from BG for U's position (e.g., by making it seem to BG that U's motivation was to prevent BG from ruining its case at the court of Western public opinion). BG's actual support, after being coerced by angry threats, was unwilling and reluctant.

The second round of dilemma elimination was not achieved at all; not, that is, at U's operational level. U negotiated with BS throughout the 10 days of the NATO ultimatum in terms of the model in table 17 and did not succeed in eliminating any dilemmas from it.

This was the model U had created by successfully bringing BG into line. In it, BS faced no dilemmas, and hence U could bring no pressure to bear on BS. Aware of this, U tried to fend off air-strikes by concessions to BS in terms of the locations to which heavy weapons must be withdrawn and the degree of their supervision, hoping thereby to entice BS into withdrawal. In this way U tried to solve its deterrence dilemma and give BS an inducement dilemma by sweetening U's offer. Apparently U failed in this, as BS simply failed to withdraw its weapons (Silber and Little, 1996, pp. 316-317).

U was saved, according to our analysis, by simultaneous action on the grand strategic level. Effectively, the politicians came to U's rescue. After British Prime Minister Major's visit to Moscow, Russian Ambassador Churkin saw the Bosnian Serbs at their headquarters in Pale, told them that Russia backed the NATO ultimatum, and offered them Russian backing if they accepted its conditions. He made this last offer by saying he would send Russian troops to monitor the withdrawal of BS weapons. Note that he did not explicitly tell BS that if they rejected NATO's conditions, then Russia would leave them in the lurch. It is undiplomatic to spell out your threats. To get willing cooperation, a positive note is necessary. In failing to tell them that Russia would, of course, back them under all circumstances, the ambassador was being sufficiently explicit and at the same time positive, as the Serbs made clear by joyously accepting his conditional offer of support.

Such was the reality. In our rerun, U takes advantage of precisely the same configuration of grand strategic dilemmas to solve the problem at operational level, although note that in doing so, it first checks its strategy with its grand strategic superiors. Checking its strategy in this way also enables it to achieve useful coordination with the grand strategic level, giving reassurance that the messages BS receives on that level will confirm what U is telling it, the essential point of which is that it cannot rely on Russian backing against NATO air strikes. Note that U's argument that BS cannot rely on R depends on the very Russian inducement and threat dilemma (that R feared a *t* that might lead to a new confrontation between East and West) that led Ambassador Churkin to visit Pale and make the same point.

In our rerun, U uses these powerful arguments to win the confrontation at operational level. BS gives in. We cannot, of course, say with certainty that U could have achieved this. BS might have persisted in its grand strategic efforts until R communicated directly with it. What we can say is that if U attempted dilemma elimination along the lines sketched in our rerun, it would have been pursuing a confrontation strategy coordinated between grand strategic and operational levels, and thereby would have been pursuing its superior's intent. It is certain, too, that such a strategy, coordinated between different levels, would have been far more effective than the ineffective and ultimately damaging efforts to sweeten its offer to BS that were pursued in reality.

In military terms, U's strategy would have seized and maintained the initiative by attacking BS's center of gravity rather than, as in reality, trying unsuccessfully to conciliate BS by progressively abandoning U's own position.

The Next Step in Our Rerun of Reality—Stabilizing the Agreement

Step 4 (A New Step 1): Assuming we have coerced BS to overtly comply with its position, U returns to the drawing board. Its new model of the situation is that seen in table 22. The dilemmas it finds in this are set out in table 24. BS and BG both have temptations to defect, preferring to break the cease-fire not only individually but collectively. They would prefer, in other words, to resume hostilities. BS would prefer this because they wish to gain ground while they can, being aware of the growing strength of a Croat–Muslim coalition being formed against them. BG would prefer it, now or in the near future, partly because they hope to be able to drive the Serbs back and partly because, if they fail, the renewed bombardment of Sarajevo might swing world opinion in favor of intervention on their behalf.

The agreement is holding for the time being because it has just been signed and UN troops have been interposed between the two sides. In these circumstances, neither side wants to be seen to be the first to break it; nevertheless, it requires stabilizing before one or the other moves against it.

Step 2: How is the agreement to be stabilized? The problem is to change the perception by BS and BG of the advantage to them of an end to the cease-fire. Unfortunately published sources give too little information for us to say how this problem can be (or was) tackled. Threatened media exposure of violations may deter BG but not BS, whose image in the media can hardly get worse. Credible threats of reprisals may be needed to deter BS. Recognizing that the dilemmas

	DETERRENCE	INDUCEMENT	THREAT	COOPERATION	TRUST	POSITIONING
UNPROFOR					√	
BOSNIAN SERBS				√	√	
BOSNIAN GOVERNMENT				√	√	

Table 24. Dilemmas facing characters in table 22.

facing BS and BG are cooperation dilemmas, for which a positive tone is appropriate, U makes sure that all threats are part of a harmonious, positive stabilization package constructed with full consultation and including carrots, however symbolic, as well as sticks.

This package must be implemented at the tactical as well as operational level. Consequently the commander's staff, after putting all the detail it can think of into an expansion of the overall model, draws up guidelines for local commanders. Within these guidelines, local commanders can analyze their confrontations and draw up local packages to eliminate dilemmas in them. As in the local commander's model in table 3, BS and U units often will take differing positions in these local models (e.g., in regard to whether certain arrangements conform to the agreement). Negative emotions then will be demanded, within the overall positive tone of the agreement.

There are potential problems of coalition relations in regard to Russian troops guarding some BS weapon cantonments. These will have to be managed in a simultaneous, coordinated manner at operational and tactical level.

Step 3: Suppose their dilemmas of cooperation and ours of trust are solved, either in reality or role-playing mode. The agreement is then stable, for now; however, it is likely that external events, interruptions, may disturb it by changing the conditions under which it holds.

Predictions Made by Drama Theory and Confrontation Analysis

Rerunning reality as we have done highlights the kind of predictions drama theory makes. It is important to understand that they are not deterministic. Neither are they probabilistic.

A character may react in various ways to the dilemmas it faces; however, if the assumptions fed in are correct, the analysis identifies the dilemmas each character must react to and allows us to predict that it will react in such a way as to try and eliminate those dilemmas. What reactions are available to it will depend on the concrete details of the particular situation (i.e., on the friction it encounters in trying to change its moment of truth in one way or another). Ways of changing the moment of truth may include changing position, changing preferences, thinking up new cards to play or taking action (e.g., military strikes) to deprive others of cards.

In this way, the theory uses models in an unusual way. We cannot, as in most other approaches, build a model, manipulate it formally to derive predictions, then come back to reality to apply those predictions. From our model-building and formal analysis, we derive not predictions, but questions about reality in these forms: What means are available for X to try and solve such-and-such dilemmas? It is our answers to these questions that provide us with the predictions we feel able to make.

Use of a confrontation model, rather than turning us away from reality, turns us back to it with searching questions to be answered.

Comments on This Rerun

This particular rerun of reality has suffered from being based on assumptions taken from third-party, retrospective, published sources, rather than the views of actual participants at the time. This affects more than the reliability of the analysis, although we have asked readers to discount inaccuracies. Drama theory is based on the game as perceived by the players. Their views, not those of others, determine their actions and reactions; therefore, theoretically it needs to be based on their views at the time.

Relying on published sources also means that we have incomplete data, particularly as regards confrontations at the tactical level. We could not say much about how BS and BG might have been brought into actual, not just apparent, compliance with a cease-fire.

Accepting all this, what difference might it have made if a confrontation strategy had been formed and used? It appears that the same result would have been achieved in bringing the BS and BG into apparent compliance with a cease-fire, with certain differences.

The bringing of BG into apparent compliance might have been done in a planned way rather than through spontaneous reactions. Further, it might have been planned to make BG more willing to comply, rather than feeling coerced.

The bringing of BS into apparent compliance might have been achieved sooner on the operational level instead of being achieved, after unavailing concessions had been made, on the grand strategic level. In achieving this result on the operational level, the commander would have been implementing his superior's intent. Some of the tactics used, in particular, the forecasting of Russian reactions, might have had to be checked with his superiors. Others would not. Coordination between the operational and grand strategic levels would have been enhanced.

Achieving the result on the operational level, in the manner outlined, would have meant some differences in the result on the grand strategic level. The actual result involved Russia. If the commander had achieved it in the manner we discuss, Russian involvement might not have been necessary. Perhaps it was desirable. Perhaps Russian involvement was a grand strategic objective. This emphasizes the necessity for a commander to check with his superior before using tactics with higher-level implications. His checking might have led to an improved plan for obtaining Russian involvement, if this was an objective.

However much or little the putative result would have differed from the actual one, it would have been achieved (if a confrontation strategy had been used, at the initiative of the commander) instead of by last-minute action at a higher level after the commander had made significant concessions from his original end state. It would have enabled the commander, from the beginning, to seize and maintain the initiative.

Insufficient data in public sources make it hard to say much about turning apparent compliance by the Bosnian parties into actual compliance. It may or may not have been achievable using available assets and within given guidelines. If it was not, attempts to formulate a confrontation strategy would have revealed this, enabling inadequacies to be reported upward and requests for extra support clearly justified.

SUMMARY OF CHAPTER 9

The analysis reported in the previous chapter is used to show how in a real-world case the commander in an OOTW might have formulated and implemented a confrontation strategy. Two rounds of dilemma eliminations that might have been carried out as part of such a strategy are suggested.

Comparing this rerun with what actually happened, the conclusion is that the first round of dilemma eliminations might have been achieved much as in reality, but in a more planned way and perhaps with greater effectiveness in inducing willing compliance. The second suggested round of dilemma eliminations was not achieved at all by the commander on the operational level; it required grand strategic action. If it had been carried out by the commander, he would have been operating a strategy coordinated between operational and grand strategic levels, seizing and maintaining the initiative, avoiding costly concessions, and carrying out his superior's intent.

While we hope this attempted rerun may be useful, it should be pointed out that it makes use of hindsight, whereas confrontation analysis is most effective when used to understand and plan for an actual, ongoing operation, not one that already has taken place. This is so because various futures that seemed only too possible at the time seem by the exercise of hindsight to have small probability just because they did not happen. This diminishes the effectiveness and impact of confrontation analysis. Use of hindsight also makes the rerun questionable, if not useless, as a test of the theory or technique.

CHAPTER 10

IMMERSIVE BRIEFING AND MEDIATION SUPPORT

The main topic of this book has been how to build and use a confrontation strategy to win an Operation Other Than War (OOTW).

In dealing with this, we have several times mentioned two related topics we now discuss in more detail.

The first is role-playing. We will discuss a method of writing briefings for role-players called "immersive briefing." This is a drama-theoretic method of looking at a set of linked confrontations from the viewpoint of each party, thinking how they must see the situation and how they must see the way each other must see it, and so forth, and basing on this a set of briefings for role-players to take the parts of real-life characters.

The second topic is mediation support. Generally we have assumed that confrontation analysis is used for unilateral decision support, helping one party deal with others. A commander would use confrontation analysis in this mode to bring external players into compliance with his end state. But, a commander also must deal with confrontations that are essentially cooperative, in that the other players are on his side and his superior's intent is that they should all coordinate their activities to achieve given objectives. Mediation between the parties is then more appropriate than support to just one player. For this we will discuss a modified form of confrontation analysis.

IMMERSIVE BRIEFING

In step 2 of a confrontation strategy, you decide on a batch of dilemmas to eliminate and think how to do it. In step 3 you carry out your elimination plan, or if you are planning rather than implementing a strategy, you think through what the results are likely to be. From there you go on to the next cycle, modeling the new situation you have brought about or imagined.

This was the process we described in chapter 7 and gave an example of in chapter 9. However, when planning or checking your strategy, you can do better than just thinking through what its results are likely to be.

There is an inherent difficulty involved. On going into a moment of truth you open up your beliefs and values to being changed; hence, it is theoretically impossible to be sure how you will come out of it. While you cannot be sure, an excellent way of stretching your imagination and letting you feel the forces of change beforehand is role-playing.

Role-playing is widely used for this purpose. Its results generally are seen by good actors as abysmal. The reason is simple: however good you are, you cannot do much with a bad script. Having a good script is not, however, a matter of having words to learn. It is a matter of knowing the life situation of the character you are acting, meaning not its personality or character, but what it is trying to achieve, and why and how, and what it thinks others are trying to achieve, and why and how. Knowing these things about your character, you can start to act, not by putting on another person's mannerisms, but by putting on its life, playing the game it has to play.

This requirement of good acting is precisely the requirement we have for exploring what may happen at a moment of truth. We need to be able to experience the moment of truth as closely as we can, and so we need to recreate the characters we will be playing against. If we can be helped to throw ourselves completely into the problems and viewpoints that other characters inhabit, we can recreate them as we want them, try out things with them, and have the added bonus of a deeper insight into their viewpoint. In this way we can check assumptions made in the course of planning a strategy (e.g., whether a non-compliant character, placed in a certain situation, has any alternative other than to give in).

Immersive briefing is a drama-theoretic way of producing briefings for role-players. Unlike most role-playing methods, it tends to produce good acting. Its effectiveness derives from its analytical basis. It puts participants into the positions of other characters, so getting them to understand their subjective feelings and perspectives and enabling them to come up with creative ideas (e.g., this character might react in such-and-such a way). Participants also come to understand their own confrontation strategy on a deep, intuitive level, enabling them to criticize it and suggest improvements.

What Comprises a Briefing

An immersive briefing is constructed using special software. With its aid you create a plot situation where a number of role-players can be briefed to interact, without scripts, as characters in a set of interconnected confrontations. There are no rules and no points to earn. The method is based on confrontation analysis of a situation, but briefings can be given to and used by role-players with no knowledge of drama theory or confrontation analysis.

Separate briefings are given to each role-player, who may be an individual or a team. A character in the confrontation that is being role-played is generally an organization (e.g., a country or a an ethnic group). Individual role-players are told to act as responsible representatives of their organization. The briefings they are given resemble those given to a commander taking over tasks from a predecessor. They should be both complete and concise. A role-player taking on a character is briefed on the following items:

- Character background, including its organization, the internal subcharacters and subconfrontations in its organization, and its relevant history

- Values and motivation, recognizing that a character will have many kinds of value systems, selfish and unselfish, long- and short-term, and that these often conflict

- Current projects (current goals)

- Current relationships with other characters

- Other characters' backgrounds, values, and projects and the relationships between them

- Confrontations it is involved in with other characters.

The logic of this briefing is that a character's background and relationships explain its values, its values explain its projects, and its projects, because they are interdependent with the projects of others, explain the confrontations it is involved in.

In relation to each confrontation, a character is told the following facts:

- Who is involved (characters)

- What each character can do about it (cards)

- What position each is taking (i.e., what solution to the conflict each recommends)

- What fallback position each character says it has in case the others will not accept its position (or it believes will not genuinely accept it)

- What future threatens if these fallback positions are implemented

- What future is forecast if currently implemented policies continue. This may or may not be the same as the threatened future

- What strategy it and the other characters currently are pursuing as each tries to get the others to accept its position.

A character may be role-played by a team of up to about five people who digest and discuss their character's briefing among themselves. During internal discussions, these teams are segregated in separate rooms, in separate corners of a large room, or even in separate locations, as when the drama is played over a computer network.

After receiving and digesting their briefings, characters (as played in role by role-players) are allowed to communicate with each other to implement their strategies (i.e., so that each character can try to get others sincerely to accept its position for implementation). For this each character tries to make credible the threats (often implicit) and promises specified in its strategy. Communication may be through arranged meetings, fixed by intermediaries or by e-mail over a network. Various communication media may be used. Communication by public announcement (e.g., giving a press conference), knowing that another character will read about it, can be simulated, such as by sending out a general e-mail purporting to be a news report of a press conference. Internal discussions continue between meetings.

Records of communications between characters help in later analysis. They can be kept automatically if communication is over a network.

What an Immersive Briefing Represents

An immersive briefing really represents a particular character's memory, or at least, the part of it that is relevant to a particular set of confrontations. If a character is an organization, its briefing represents its organizational memory. That is why a briefing may be likened to the briefing given a new commander when he takes over responsibilities from a predecessor.

Three things follow from the fact that a briefing represents a character's memory:

• A character can consult its briefing at any time.

• As a situation develops into a new one, a briefing becomes out-of-date. Like a diary that stops at a certain date, it records things as they were, or seemed to be, at a point now in the past. Replacement of the briefing by a more up-to-date one is possible. Alternatively, characters can be left to make their own notes of new developments.

• Briefings are subjective and differ from one another. Character A may be briefed differently from B about past events, about its own or another's position in a particular conflict, and so forth. Certainly characters' briefings will differ in the values reflected in them, because each briefing will reflect the values of the character being briefed. Consequently briefings are second- and third-order subjective, as well as first-order; that is, characters are briefed as to (their beliefs) about others' beliefs and values, others' beliefs about their own and others' beliefs, and so forth (see Bennett, 1977).

Using Software

The package for authoring an immersive briefing produces a suite of software briefings, one for each character, reflecting the characters' different viewpoints and information. Current software provides briefing information both in text form and card-table form. Text or card-table is called up in a computer screen window by clicking on menu items or pictures of the characters and the confrontations between them. Further clicking on different parts of a card-table calls up textual information about characters, positions, fallbacks, and strategies in that confrontation. Figure 3 shows how a briefing written for the character, "Bosnian Govt," looks before any pictures or menu items have been clicked. This briefing was one of an experimental suite of briefings written for DERA.

Pointing and clicking through an immersive software briefing is, even without role-playing, an effective way of absorbing a confrontation analysis. It might also be used, therefore, to present and distribute analyses. Users could brief themselves by looking through the briefings given to different characters and comparing their viewpoints.

Friction and Change in Immersive Role-Play

Immersive briefings generally are designed to bring characters to a moment of truth. Recall what this means. A moment of truth is a point at the climax of a confrontation when characters achieve the following:

- They finally understand each other (or think they do), including each others' positions and fallback positions.

- They are, as a result, brought up against the need to change their fixed views of themselves, their situation, and each other if they are to avoid falling into conflict.

At such a point characters tend to change their definition of the confrontation they are in by redefining the characters in it, the cards they can play, and their preferences between futures, as well as their positions; however, their ability to do so is limited by the weight of existing values, evidence, mission definition, and

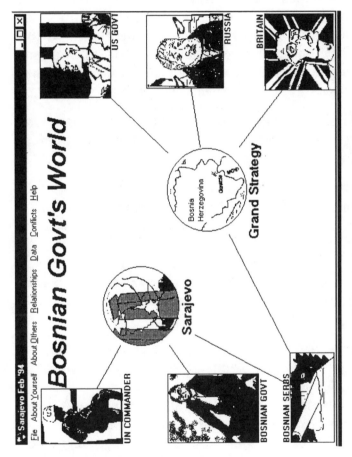

Figure 3. Computer screen showing briefing
on character role for Bosnian government.

so forth (i.e., by friction). Emotion enables them to break out of existing assumptions and routine reactions, and so to overcome friction.

Role-players accordingly are told that the frames and positions their briefings give them are initial frames and positions only. They have inherited them from their predecessors but they can change them. In this respect, immersive role-playing differs from game-playing or ordinary role-playing. The only constraints are imposed by the following circumstances:

- The past, consisting of characters' backgrounds and the events and projects that brought them to their current impasse

- Characters' expectations of each other, and the difficulty they may find in changing these. It will be of little use to me to undergo a conversion so that I prefer to carry out a promise or threat if you continue to believe I will not. Equally, a change of position or discovery of a new option not listed in the card-table will need to be communicated convincingly

- Role-players' sense of responsibility for properly representing their character. An immersive role-play can, in theory, be spoiled by an irresponsible player who acts out of role. At the other extreme, role-players may need to be reminded that they are free to change, as some are over-responsible. Such reminders can be given in debriefing sessions arranged to follow each round of negotiations.

Over-responsibility for the reality of your character is a particular problem in role-playing past real-life events, as players tend to think they should do as their real-life counterparts did; therefore, it is best to fictionalize past events, at least to the extent of changing characters' names.

The degree of friction a character meets should depend on its briefing, which tells it how important certain values are to it and which facts it sees as incontrovertible, and on the pressures brought to bear on it in interactions with others. Emotions engendered in a role-player may enable it to overcome even considerable friction to reframe the situation. Role-players should experience the fundamental paradox of drama theory (i.e., the more I feel hemmed in by fixed constraints, the more emotion I feel, hence the more I am inspired to think of changes, so that the more unchangeable a situation feels, the more changeable it becomes). The realism of this experience is enhanced if players first run a role-play under instructions that they should stick strictly to their briefings. Having seen and discussed the result of this, they may then replay the scenario with instructions giving them more liberty to innovate.

Immersive Role-Playing as a Means of Prediction

Suppose an immersive role-play is based on accurate analysis of a real-world situation. Then we study the forces at work in the real world by studying role-player behavior. This can yield predictions of a sort, as when we find that a character seems to have no alternative other than to comply, or vice versa, when we find that there are no pressures on it to do so.

We cannot reliably get such predictions simply by running a role-play and observing the results. Any prediction of this sort should be subjected to criticism and discussion aimed at overthrowing it. The role-play itself should be used to draw attention to questionable assumptions.

The point is that a drama-theoretic prediction, if based on the presumption that there is only one way to overcome a dilemma, such as by complying with our position, itself generates pressures for its overthrow because it motivates the character concerned to find some other way. The only pure predictions generated, even if our model and briefings are accurate, are to the effect that characters will face and attempt to eliminate certain dilemmas. Further predictions are obtained by asking: "What kinds of dilemma elimination are likely to succeed?" Answering this question depends on estimating the friction players will meet in different directions, hence any useful predictions depend on common-sense estimates of friction.

How Immersive Role-Playing Helps

How, then, could immersive methods help a commander develop a confrontation strategy? We suggest in the following ways:

- **Foreseeing likely reactions**—Role-players' behavior can be used to foresee possible real-world reactions. Firm predictions of this kind should be subjected to criticism of assumptions before being accepted.

- **Encouraging criticism of key assumptions**— Dilemmas can inspire role-players not only to change their briefings but to criticize the assumptions made in them.

- **Asking key questions**—Assumptions made about beliefs and preferences in the briefings given to role-players are key to the development of a confrontation. Questioning them raises key questions for a commander to answer.

- **Investigating assumptions**—Assumptions exposed to criticism can be investigated both theoretically (i.e., asking "What dilemmas are strengthened or weakened if this assumption is varied?") and in real life (i.e., intelligence can be directed to answering specific questions about other parties' beliefs or preferences).

- **Understanding the implications of a confrontation analysis**—A complex analysis can be hard to digest. Role-playing gives intuitive understanding.

- **Understanding intuitively the gradient of a moment of truth**—The term "gradient" is used by analogy with a physical system, where it is an object showing the directions a system is tending to move. The drama-theoretic gradient of a moment of truth is an object that sums up its tendency to change as characters attempt to eliminate dilemmas. Intuitive understanding of the gradient of a confrontation is what a commander needs to orchestrate the entire compass of beliefs, attitudes, and emotions to tilt the gradient toward his position.

- **Rehearsing real-world interactions**—We perform better in a situation after having developed skills and intuition through rehearsal.

- **Enhancing unity of effort across different commands and levels**—Joint role-playing by those responsible for operations at different commands and levels can generate common understanding of the confrontation, enhancing unity of effort.

- **Appreciating the viewpoints of other participants**—An immersive briefing does not take a single viewpoint, nor even a neutral one. It takes each viewpoint and its viewpoint on each other's viewpoint and…so forth; thus it provides a conceptual platform from which to survey all the viewpoints.

MEDIATION SUPPORT

Unilateral Decision Support—General Considerations

Most of our discussion of how to win an OOTW has assumed that the requirement is to support the decisions of one character, ourselves, in confrontation with others. In other words, we have been assuming that our confrontation analysis is in unilateral mode.

What does this imply? First, that all information we use is collected and kept by ourselves, except when we selectively decide to release it to others. For example, we may construct, for role-playing purposes, a version of each character's viewpoint and its viewpoint on others' viewpoints, and so forth; however, all these viewpoints are based on our own information, we do not go directly to other characters and ask for their viewpoints.

Second, unilateralism implies that our strategy is similarly confidential. We tell others only those parts of it we wish them to know. They know the messages we send them, or should know them. Messages sent by physical action, such as bombing or delivery of aid, should be accompanied by clear explanations lest they be misinterpreted and send a message that was not intended. Such clear explanations do not amount to a full disclosure of our confrontation strategy.

Unilateral mode is the default mode for players that neither belong to a common organization nor have an acknowledged common purpose. Each must then have a strategy for pursuing its ends and must send messages to others as part of its strategy. Its ends may be selfish or benevolent, atomistic or communal, but they are its own ends.

However, we saw in chapter 3 that the very process of conflict resolution between separate parties tends to generate, through the emotions and arguments they develop in trying to influence each other, a common purpose and a common organization to support that purpose. We sometimes express this by the phrase, "every drama is a character." This is how nations develop and private revenge is replaced by public justice.

The development of law is only one example (although an extreme and important one) of a general process by which organizations develop formal and informal procedures for conflict resolution. The essential characteristic these procedures share is that they are public and open, dependent on every step taken in them being common knowledge among the players, whereas the formation and implementation of a confrontation strategy is essentially private. Even telling other characters my confrontation strategy does not make it common knowledge because saying that it is my strategy may be merely part of my strategy. I may be lying.

By contrast, many structures and procedures of an organization must be public or they defeat their purpose. It makes no sense to have a secret capital city or a secret constitution. Many of these public organizational procedures have an important conflict-resolution function. This is obviously true of arbitration and appeal procedures. Command procedures resolve conflicts by specifying which officer has to make a decision when there are conflicting views.

Informal conflict-resolution procedures are more tricky. Suppose an office makes decisions by consensus, but a new recruit does not know that and expects the manager to decide on his own. This example illustrates both the need for procedures to be common knowledge if they are to be effective and the fact that informal procedures are free to change and develop with the players that operate them.

It is important to realize that each character, in choosing how to go through the public steps of an open conflict resolution process, is still pursuing its own, private, intuitive confrontation strategy. (Here by "intuitive," we mean "naturally developed, not derived from knowledge of drama theory or confrontation-analysis methods.") Realizing this, we see that the unilateral mode of implementing a private confrontation strategy is not replaced by open institutions for conflict resolution. It continues to be pursued within them. The difference they make is that each private strategist's choice of cards to think up and play is constrained by the rules of the institution it is playing within, because to break these rules would sometimes incur punishment, at other times would be self-punishing in diminishing the impact of the common-interest arguments the strategist is using to make others accept its position.

Thus, unilateral mode continues universally, even within conflict-resolution procedures; however, it is often seen as a Machiavellian pursuit of selfish interests to the detriment of the common purpose. Why?

We all have a common interest in strengthening the conflict-resolving institutions in our society and influencing them to develop in ways that advance common ends. Clear, intelligent development of private confrontation strategies is thought to endanger these institutions by showing how to take advantage of them.

A generally excellent book on how to negotiate, Fisher and Ury's *Getting to Yes* (Fisher and Ury, 1983), heads its first chapter, "Don't Bargain Over Positions." This would seem to contradict the whole of drama theory. On examination, it does not. To begin with, the authors recommend that each negotiator should solve its deterrence dilemma, if any (i.e., have a fallback position that puts pressure on the other). If the situation is not to escalate, it follows that they must solve their inducement dilemmas by converging to a common position. The authors implicitly assume this. Given it, the essence of their advice is, "Don't talk about positions, use rational arguments in the common interest." Within the given structure, that is excellent drama-theoretic advice. Not talking about positions does not mean they do not exist. Rational arguments in the common interest depend on the fact that for each negotiator, even the other's position is preferable to the threatened future, avoiding which is a priority for both. The arguments depend on finding the common interests responsible; hence, the arguments implicitly but clearly refer to positions.

In general, the animus against private strategies leads to much hypocrisy. It is as if unilateral mode is seen as part of our sinful nature. It cannot be avoided or denied, but must be discouraged by being starved of intelligent development.

We suggest that this attitude is partly correct. Unintelligent, cynical expositions of how to form and implement private strategies do undermine societal and organizational conflict-resolving institutions; however, drama theory shows how these very institutions develop out of the pursuit of private strategies, as private strategists find a need to argue for the common interest. Drama-theoretic confrontation strategies should strengthen, not undermine, the institutions.

Meanwhile, the animus against private strategies exists. It means that unilateral mode often will be seen as appropriate only when parties are not part of a common organization or are not pursuing an acknowledged common purpose. Even then, it will seem appropriate only if we approve of the purposes our side is pursuing. Fortunately, these conditions are met if we limit ourselves, as we have largely done, to confrontations between our defense forces and potential rebels against the New World Order. This is the post-Cold War world described in chapter 1.

Mediation Mode

We now ask, "Can we use drama-theoretic insights into the private strategies pursued within open conflict-resolving procedures to improve the conflict-resolving procedures themselves?"

The answer is that we can. Confrontation mediation (i.e., doing confrontation analysis in mediation mode) involves doing a confrontation analysis that is seen by and is common knowledge for all characters. The aim is not to help one side to win, but to help the process of conflict resolution itself. We want to move it in a positive direction, toward a cooperative, happy ending, rather than negatively, toward a tragic one.

The following question arises: Knowing that parties are pursuing private strategies, how can we trust the information they submit to an open, common-knowledge confrontation analysis? We will answer this and other questions after outlining the procedure, giving a simple example to make the discussion concrete.

How Confrontation Mediation Is Done

Step One—Analysts go to each side and interview those involved. Views are solicited and carefully noted, without criticism. This is important, not only to build a good model but so that participants realize their views have been incorporated (i.e., the model is not biased against them).

Step Two—The confrontation is analyzed using data from Step 1. A model is built. There are now two cases, as follows:

- *The model shows full and harmonious agreement.* It is presented to the parties. If they confirm the agreement, the process ends; otherwise, it goes to Step 3.

- *The model contains dilemmas.* It is presented to the parties, perhaps through immersive role-playing with fictional names substituted for actual ones. In discussions, dilemmas are high-lighted and dilemma elimination encouraged through positive discussion and arguments in the common interest.

Step Three—Parties confer among themselves to see if they can confirm with their internal subcharacters any changes in attitude or beliefs generated at Step 2. The process then goes back to Step 1.

For example, suppose a memorandum of understanding must be negotiated to provide for cooperation between two forces, ourselves and another. Our team at the negotiations (call them representatives) want to make commitments that those responsible for fulfilling the commitments (call them suppliers) see as extreme, but representatives see as necessary. Each side, representatives and suppliers, has a reasonable case, but sees the other as unreasonable. Each side is interviewed.

Step One—In interviews, representatives give excellent reasons for the need for the commitment, suppliers for the uncertainty of being able to fulfill it. Neither side will budge.

Step Two—The simple model in table 25 (first three columns) is built and presented to the parties. After role-playing they converge to the compromise in column C. Representatives agree not to make the commitment formally because suppliers are unsure they can fulfill it, while suppliers agree to make every effort to fulfill it, although they cannot guarantee success.

Step Three—Each side confirms its changed position internally.

Step One, round two—Interviews reveal that each side means to keep the agreement, but distrusts the other.

Step Two, round two—The new model is presented. Each side's trust dilemma is discussed and eliminated.

Step Three, round three—Newly acquired trust in the other side is confirmed internally, reported on return to **Step One, round three,** and reconfirmed by each party to the other at **Step Two, round three.** End of process.

Harnessing Private Strategies to the Mediation Process

Let us now answer the question, "How can model-builders in this process trust the information they are given in interviews, knowing that each party will be trying to influence the process in its favor?"

By choosing the right party to trust for each piece of information, a mediation model can be made deception-proof in the following sense: each party's attempts to influence the process in its favor will help the process rather than hinder it.

To show how this works, we will use as an example the process by which mediation resolved the simple confrontation between representatives and suppliers in table 25. In the first model, constructed in the first Step 2, each side has an inducement dilemma (preferring the other's position to *t*) and a threat dilemma (preferring not to carry out its threat, if the other does).

Now observe that it is in neither side's interest to admit that it faces these dilemmas; therefore, if either of them refuses to admit to a dilemma, we can point out that what matters to it is not, actually, whether it has the dilemma, but whether the other believes that it has. Seeing this, it can be encouraged to produce rational common-interest arguments and give evidence to convince the other it does not have the dilemma. Its use of common-interest arguments in particular can be encouraged by pointing out their greater effectiveness in influencing the other's beliefs, which is what it needs to do.

If, on the other hand, either party cannot believe that the other is unconvinced, the purportedly unconvinced party is shown that what matters to it, again, is not that it is unconvinced, but that the other should believe that it is; therefore, it can be encouraged to use rational common-interest arguments and produce evidence as to why it ought to remain unconvinced.

This general principle, of laying the burden of producing conviction on the one who needs to do so, and suggesting how it should be produced, will, if followed consistently, bring about convergence to common beliefs through the use of reason and evidence, passionately argued.

	R	S	*t*	C
REPRESENTATIVES	1	3	4	2
Make commitment	■	□	■	□
SUPPLIERS	3	1	4	2
Fulfill commitment	■	□	□	■

LEGEND
■ means card is played
□ means card is not played
R is Representatives' position
S is Suppliers' position
C is Compromise, in which R doesn't commit but S tries hard to fulfill
t is threatened future
# indicates preference ranking (1 is most preferred)

Table 25. Confrontation between representatives and suppliers.

Moreover, in this process of convergence, characters will have made appeals to and explorations of their common interests. Their motivation will have been to convince each other, but the effect will be to construct a view of themselves as having interests in common that will prompt them to think of win-win solutions and compromises.

Continuing with our example, suppose that representatives and suppliers have found a compromise solution and in Step 2, round two, found that they cannot trust each other. Each is told that its problem is not whether it is trustworthy, but that the other should believe that it is. As before, it needs to produce rational common-interest arguments and evidence to convince the other, and may in the process convince itself, if it had in fact intended to defect, that defection is against its interests.

In this way, the process again brings about convergence to common beliefs with accompanying reframings of the situation that actually may change perceived facts, as well as beliefs. The general principle to be observed is that the unbelievability, for good structural reasons, of a character's evidence can be used in the mediation process to persuade that character to provide better arguments and evidence, based on the parties' common interests, simply because it is always other parties, not itself or the mediator, that it needs to convince.

When Should It Be Used?

Theoretically confrontation mediation is usable whenever parties are willing to participate; however, it must be done by external experts, who require funding. The party that funds these experts naturally lays itself open to the suspicion that it is pursuing a private strategy. Such suspicions are certainly aroused by attempted mediation in many OOTW confrontations.

To overcome this problem in a fundamental manner, confrontation mediation, like established conflict-resolution procedures such as legal systems, needs to be funded by an organization to which all parties belong and owe loyalty.

This is because the above suspicion is justified. The funding organization must be pursuing a private strategy; we all are. The question is, with what aim? The only answer that will satisfy the parties is the general, long-term, common interest of the parties themselves. To ensure this aim requires an institution such as the law, that is owned by an organization (here, the nation) to which all parties belong, yet is kept separate from its policy institutions (because policies may be what the confrontation is about).

Arrangements like these can be made within an organization such as a firm or government agency. A mediation unit can be set up with guarantees of independence. It might even be done within a coalition of defense forces assembled for a specific OOTW. Our argument is that unless it is done, confrontation mediation will be used at most sporadically.

Informal Use—Being Your Own Mediator

When formal mediation is impracticable, the ideas of mediation may be of use to those implementing their own, private confrontation strategies.

There is a sense in which I need to be my own mediator. If I wish to avoid tragic outcomes, as by definition I should, then I want the process as a whole to go in a positive direction, in addition to wanting it to go in my own particular direction. I can help this to happen by observing myself and other parties in mediation mode and giving mediation-type advice to myself and them as needed. Familiarity with the principles of mediation support is useful to all involved in confrontations.

Another consideration is that confrontations vary along a continuum of expected cooperativeness according to the degree to which cooperation is expected or demanded of them. This is not a matter of how characters feel. Often parties in the most hopefully cooperative relationships have the strongest negative feelings, just because they cannot cooperate. Vice versa, parties trying to kill each other may feel very little toward each other, or may even feel friendly. Emotions, as we have seen, have particular functions at a moment of truth within a confrontation; they do not make confrontations cooperative or not.

That is brought about by organizational relationships. Parties are expected to cooperate when they are in functional relationships within the same or related organizations. Hostile armies confront each other non-cooperatively in our sense, whereas armies in a coalition, who may frustrate and infuriate each other as much or more, do so even though their confrontations are hopefully cooperative.

It is this variable factor of expected cooperativeness that seems to determine how strongly private confrontation strategies are deplored, regarded as disruptive, and denied funding. Organizations deplore Machiavellian behavior within their ranks. We might hope to use mediation support instead of unilateral confrontation support to deal with such problems, except that entrenched positions and angry attitudes often found in hopefully cooperative confrontations may make it hard to get participation. In such cases, being my own mediator may be the best solution available.

SUMMARY OF CHAPTER 10

This chapter deals with two topics not covered elsewhere in the book. Immersive briefing is a drama-theoretic tool for briefing role-players to enable them to take the parts of characters in a confrontation that has been analyzed. Each role-player is told about its character's background, values, projects, relationships with other characters, and confrontations with others. In relation to each of its confrontations, it is told who the characters involved are, what their positions and fallback positions are, and which tactics they are using. Each character in a drama gets its own briefing, different from the others. A character may be role-played by a team or an individual.

Briefings are composed using a software tool and distributed as software packages, through which information is accessed by clicking and pointing. After being briefed, role-players interact with each other to try to resolve their problem. Interactions can be via various media, from e-mail to personal meetings.

Immersive role-playing can help a commander to develop a confrontation strategy by enabling him at step 2 to test out the likely effects of his strategy on other players. The results of such testing need to be thoroughly criticized and discussed.

Mediation support is contrasted with unilateral decision support as an alternative mode of using confrontation analysis. Instead of being used to develop a private strategy for one player, confrontation analysis in this mode is used to model a problem for the equal benefit of all players. Information is obtained from all and distributed to all. The modeling process is used to help the conflict resolution process move toward a happy ending rather than a tragic one.

It might seem problematic to rely on the information supplied by each participant in confrontational mediation because each will want to give information that will bias the procedure toward its own ends; however, this is not a problem because this private motivation of each participant can be used to drive the process forward. This is done by pointing out to a participant whose information is doubted that it is not the truth of its information so much as its acceptance by others that matters to the participant. In this way, the participant is encouraged to use rational arguments in the common interest to convince others.

Mediation mode may be considered most appropriate to problems where parties belong to a common organization or have acknowledged common ends, whereas unilateral mode may seem right only when these conditions are not met; however, unilateral mode also encourages happy endings, if used intelligently.

Appendix

The Mathematics of Drama Theory

Introduction

D rama theory has an essential mathematical foundation. This appendix reviews the basic mathematical facts involved. It generalizes the treatment of Howard (1998), as it deals with general positions involving sets of individual proposals rather than singular ones.

First, a warning about the use of mathematics is particularly relevant to drama theory. Use of mathematical notation can give the impression that only the abstract structures captured in the mathematics matter. For example, use of a mathematical model that is symmetric between characters may seem to assume that characters have symmetric roles. This does not follow, any more than to measure two people and find they have the same height is to assert their identity. Because the mathematics to be presented captures essential structures underlying a situation, we must emphasize the importance of all the details not captured. They are important in using and interpreting the mathematical structures themselves.

The Frame

A *frame* (the drama-theoretic equivalent of a game, modeled by a card-table) may be defined as a pair,

$$(h, u)$$

where the following points apply:

- $h:D{\rightarrow}C$ is a function assigning *cards* in a deck D to *characters* in a *cast* C

- hd is the character that controls the card d

- h^* is the inverse function from subsets of C to subsets of D. Thus, h^*G (where G is a subset of C) is the set of cards controlled by members of the group (subset) of characters G

- Subsets of D (sets of cards) are called *selections*. A selection is formed when a group of characters G chooses (plays) a subset from the set h^*G of cards it controls. A single character c makes a single selection from its hand $h^*\{c\}$.

- When all characters make a selection, the result is called a *total* selection.

- $u=(u_c|c{\in}C)$ is a family of *utility* functions, one for each character c in the cast C. Here, the utility function of c is a function $u_c:S{\rightarrow}R$, where R is the set of real numbers and S is the set of selections (subsets of D). The number $u_c(s)$, where s is a selection in S, is the utility character c gets from s. Although a real-valued function, u_c is regarded as having ordinal significance only (i.e., any function that ordered outcomes in the same way as u_c could be used in its place).

Note that there is no reason to limit consideration to frames that are finite, although later we find a use for the assumption that the number of distinct positions is finite; however, both characters and cards can be infinite in number. If each hand is countable, each character selects from a copy of the real line. Then if we have an infinite cast, S is an infinite-dimensional Euclidean space.

Notation: For any selection s and *group* $G \subseteq C$, write

$$s_G$$

for

$$s \cap h^*G$$

This is the subset of s played by members of G.

Write

$$s >_c t, s >_G t$$

to mean, respectively,

$$u_c(s) > u_c(t) \text{ and } u_c(s) > u_c(t) \text{ for all } c \in G$$

The symbol $\geq$ is used similarly.

General Positions and Moments of Truth

To communicate, characters develop a shared frame called a *common reference frame*. Within this frame they take up positions by specifying which cards in any player's hand they insist should be included in a final resolution of their joint decision problem and which should be excluded. This means that each character chooses as its position a non-empty subset p of the set S of selections with the following characteristic:

$$\exists t, t' \in S : p = \{s \mid t \subseteq s \subseteq t'\}$$

A set p with this characteristic will be called *closed*. The set t will be called the *lower bound* of p and the set t' its *upper bound*. When p is a character's position, the lower bound t will be that character's set of *included* cards and the complement of the upper bound, the set $D-t'$, will be its set of *excluded* cards.

Closure simply means that the set can be represented by a single column of the card-table, each cell of which contains either a card (one the character insists be included), the absence of a card (one that it excludes) or the symbol $\sim$ (for a card it takes no position on).

Theorem 1—Any intersection of closed sets is closed.

Proof—We construct the intersection of a set of closed sets as a closed set whose lower bound is the union of their lower bounds and its upper bound, the intersection of their upper bounds. ∎

We call the selections belonging to a character's position its *proposals*.

In addition to its position, each character c chooses a *fallback position*, a selection from its own hand, $h^*\{c\}$.

When characters see their positions and fallback positions as *final*, they arrive at a *moment of truth*. This is a triple (F,p,f), where:

• F is their *common reference frame*

- $p=(p^c|c \in C)$ is a family of non-empty, closed sets of selections, p^c being the position of c

- f is a single selection called the *fallback*. The fallback position for each c is $f_{\{c\}}$

- The deck of cards D is minimal relative to the positions of p. This means that it contains only cards that are either included in some character's position or excluded by some character's position. This models the fact that at a moment of truth, characters will simplify matters by excluding from consideration cards that none of them takes a position on, and therefore which are not an issue for them.

The following terminology and notation are used concerning a non-italic *moment of truth* (F,p,f):

- For any group $G \subseteq C$, write p^G for the intersection $\cap(p^c|c \in G)$, and call this the *position of group G*

- If p^G is non-empty, call the group G *compatible*

- For any G, define the group $OG=\{c|p^c \cap p^G=\varnothing\}$ and call it the *opposition* to G. It consists of the characters whose positions are incompatible with that of G.

Note that each group position is closed by theorem 1, being an intersection of individual positions. Also, if G is incompatible ($p^G=\varnothing$), then $OG=C$. Everyone opposes G. If $G=\varnothing$, then $p^G=S$ and $OG=\varnothing$. No one opposes G.

Theorem 2—The position p^c of the whole cast is either empty or singular (i.e., it contains just one proposal).

Proof—Suppose p^c is non-empty. Because the deck D is minimal with respect to p, every card is either included in some character's position, in which case it is in every proposal belonging to p^c, or excluded by some character's position, in which case it is in no proposal belonging to p^c. No card can be both (i.e., included in some character's position and excluded by another's), or p^c would be empty. Hence we have determined, concerning any card, whether or not that card is in a proposal belonging to p^c (i.e., we have determined a unique proposal belonging to p^c). ■

Now we propose to show that at a moment of truth characters find they have reached a completely satisfactory resolution of their joint decision problem or they face dilemmas. First we must define *completely satisfactory resolution* and *dilemmas*.

Strict, Strong Equilibria

Call $T \subseteq S$ a *strict, strong equilibrium* (SSE) if it is closed and no group G has a *potential improvement* from it, where the set $Imp_G(T)$ of *potential improvements for G from T* is defined as:

$$Imp_G(T) = \{s \notin T | \exists t \in T : s_{-G} = t_{-G}; s \geq_G t\}$$

Thus a potential improvement from T is a selection outside T to which a group G can move unilaterally (i.e., given that those not in G do not change their selections) from a selection inside T without any member of G losing utility from the move.

Accordingly, a SSE is such that no group *G* can move unilaterally from a selection in it to a selection outside it without loss of utility to some member. Hence if all characters in the cast *C* agree to implement a SSE *T*, each individual or group within *C* must mean to do so, because the following points apply:

- *T*, being closed, has the characteristic that if each character *c* independently chooses a selection from its hand *h*{c}* that is compatible with *T* (i.e., contains all the cards in *h*{c}* that are in the lower bound of *T* and none that are not in its upper bound), then the total selection will belong to *T*.

- As a consequence, any group *G* that plans to break the agreement while expecting those not in *G* to keep it must expect to move from a point in *T* to a point outside it, which means that at least one of them must expect to lose.

Because of these characteristics, a strict, strong equilibrium that is the position of the whole cast is a *completely satisfactory resolution*. It is satisfactory to all of them in that it is compatible with all their positions and, furthermore, no group can be suspected of intending to defect from it. Finally, being the position of the whole cast, it is singular (theorem 2). Therefore it represents full and complete dramatic resolution.

Theorem 3—Any intersection of strict, strong equilibria is a strict, strong equilibrium.

Proof—First, such an intersection is closed by theorem 1. Next, a potential improvement from an intersection of SSE would be a point *s* not belonging to at least one of those SSE that is at least as good for all members of some group *G* as a point *t* belonging to every SSE, and is reachable by *G* from *t* (i.e., *G* can move to it from *t*). Such an *s* would be a potential improvement for *G* from an SSE to which it does not belong; but there are no potential improvements from an SSE, proving the theorem. ∎

The dilemmas characters face at a moment of truth may now be examined by defining, for a given character *c*, six gradient sets, the non-emptiness of each of which puts *c* in a specific dilemma.

What Is a Gradient?

The *gradient* of a moment of truth is its tendency to cause characters to reframe their situation. This tendency arises from the non-emptiness of certain sets representing dilemmas facing characters. We use the term in analogy with the gradient of a dynamic system, which is an object that represents the system's tendency to change in various directions.

Dilemmas motivate characters to attempt to change (redefine) their moment of truth. There is a gradient set for each character and each type of dilemma. If and only if all gradient sets are empty for all characters, the gradient of the moment of truth as a whole is zero and there is full and complete dramatic resolution.

The gradient sets for a given character are the *deterrence, inducement, threat, positioning, cooperation*, and *trust* gradients.

Cooperation Gradient—The *cooperation gradient* for *c* is the set of potential improvements upon the position of *c* for groups containing *c*, written as follows:

CoGrad(*c*)=
$$\cup\{Imp_G(p^c)|c\in G\}=\{s\notin p^c|\exists t\in p^c{:}\exists G{:}c\in G;s_{-G}=t_{-G};s\geq_G t\}$$

If this is empty, *c* is said to be *trustworthy*; otherwise, *c* is untrustworthy in relation to at least one of its proposals, in that in the Implementation phase (see figure 2), it might be tempted not to carry it out. Note that if *c* does not need to offer this proposal, *c* can become trustworthy by eliminating it from its position; otherwise, the untrustworthiness of *c* is a serious matter.

Trust Gradient—The *trust gradient* for *c* is written as:

TruGrad(*c*)=
$$\cup\{Imp_G(p^c)|c\notin G\}=\{s\notin p^c|\exists t\in p^c{:}\exists G{:}c\notin G;s_{-G}=t_{-G};s\geq_G t\}$$

If it is empty, *c* is said to be *trustful*; otherwise, it has to be mistrustful in relation to at least one of its proposals, in that in the Implementation phase, others might be tempted not to carry it out. If this proposal is not important to *c*, *c* can become trustful simply by eliminating it from its position.

Theorem 4—**If a character is trustworthy and trustful, its position is a strict, strong equilibrium. Hence, if a cast is trustworthy and trustful, then the intersection of all its positions is a strict, strong equilibrium.**

Proof—If both gradients are empty for *c*, there are no potential improvements from p^c. The rest follows from theorem 2. ∎

Deterrence Gradient—The deterrence gradient for c is now defined as the set of all selections that are (a) reachable from the fallback f by some subgroup of those opposed to a group that contains c, and, except for the presence of c in it, would have no opposition; (b) preferred by all of them to every proposal in that group's position. It is written:

DeGrad(c)=
$$\{t | \exists G, H: OG \supseteq H \neq \varnothing; O(G\text{-}\{c\}) = \varnothing; t_{-H} = f_{-H}; t >_{H} p^{G}\}$$

If **DeGrad**(c) is non-empty, the demands of c (i.e., the position c takes) mean that it belongs to a compatible group whose fallback position puts no pressure on a subset of the opposition to accept its position. In fact, it pressures that subset not to. Thus c is in a dilemma: it must either adopt a position incompatible with this group (which may be a one-person group consisting of c itself) or else find some way of putting pressure on the opposing subset. If **DeGrad**(c) is empty, c can be called *realistic*: its position does not cause it to belong to any group that does not pressure all its opposition to accept it.

Inducement Gradient—The *inducement gradient* for c is the set of proposals in positions opposed by c that give c at least as much utility as the fallback f.

$$\textbf{InGrad}(c) = \cup \{p^{G} | c \in OG\} \cap \{s | s \geq_{c} f\}$$

If non-empty, it would be rational for c to accept a proposal in this set, and so cease to oppose the position containing it, rather than suffer f, yet c is insisting that it will not do that. Thus an $s \in$ **Ingrad**(c) offers an inducement for c to shift or enlarge its position. If **InGrad**(c) is empty, c is said to be *uninducible*.

Theorem 5—A cast that is realistic and uninducible is compatible (i.e., at least one proposal belongs to all cast positions) provided the number of distinct positions they hold is finite.

Lemma—If C is incompatible and its members take a finite number of positions, some compatible, opposed subgroup of C contains a member c without whom it would be unopposed.

Proof of lemma—Partition C into groups of characters holding the same position. Because the number of positions is finite, this gives a finite partition, each group of which is compatible. Order these groups in an arbitrary order. If group one is unopposed, add the second group to it (i.e., form their union), obtaining a compatible group because the second group did not oppose the first. If the union of first and second groups is unopposed, add the third, and so on until either a compatible, opposed group is formed or all groups have been added together to form a compatible group. But that is impossible, because C is incompatible. Hence there must be a group i, the addition of which to $\cup$ (group one, group two,… group [i–1]) first created a compatible, opposed group. Let c be any member of group i. The addition of c to $\cup$ (group one, group two,…group [i–1]) creates a compatible, opposed group that would be unopposed without c. ■

Proof of theorem 5—Suppose every character in C is realistic and uninducible and the number of its positions is finite. Now suppose, if possible, that C is incompatible. By the above lemma, some character c belongs to a compatible, opposed group G such that

$O(G\text{-}\{c\})=\varnothing$. Because c is realistic, **DeGrad**(c) is empty, so that

$$\forall t_{OG}:\exists b\in OG, s\in p^{G}:t_{OG}\cup f_{\text{-}OG}\leq_{b}s.$$

In particular, letting $t_{OG}=f_{OG}$,

$$\exists b\in OG, s\in p^{G}:f\leq_{b}s.$$

Such an s is in $\cup p\text{-}p^{b}$. Therefore, for some b, **InGrad**(b) is non-empty, contrary to our assumption. ∎

***Theorem 6 (theorem of the final state)*—A cast that is trustful, trustworthy, realistic, and uninducible is compatible at a joint position that is a singular SSE, provided the cast takes a finite number of positions.**

Proof—Suppose a cast is realistic and uninducible (and takes a finite number of positions). From theorem 5, its positions intersect. From theorem 2, their intersection is singular. If all its members are trustful and trustworthy, that intersection is a SSE (theorem 4). ∎

Positioning Gradient—The *positioning gradient* for character c is the set of all proposals preferred by c to some of its own proposals, even though they belong to positions c opposes. These pose a dilemma for c, inasmuch as c will find it hard to argue against positions, some proposals in which it likes better than some proposals of its own. This gradient set is written

$$\textbf{PoGrad}(c)=\cup(p^{G}|c\in OG)\cap\{s|\exists t\in p^{c}:s>_{c}t\}$$

Theorem 7—If a cast is compatible, all positioning gradients are empty.

Proof—In a compatible cast, no set *OG* has any members. ■

Threat gradient—The threat gradient for *c* is written as:

$$\mathbf{ThGrad}(c)=\cup\{Imp_G(\{f\})|c\in G\}=\{s|\exists G\ni c: s_{-G}=f_{-G}, s\geq_G f\}$$

It contains potential improvements from *f* for groups *G* containing *c*. Thus it shows how the threat of *c* to implement *f* may lack credibility because in the Implementation phase *c* might be tempted not to carry it out.

BIBLIOGRAPHY

Alberts, D.S., & R. Hayes. *Command Arrangements for Peace Operations.* Washington, DC: National Defense University Press. 1995.

Bennett, P.G. & Howard, N. "Rationality, Emotion and Preference Change: Drama-Theoretic Models of Choice." *European Journal of Operational Research.* 92: 1996. 603–614.

Bennett, P.G. "Toward a Theory of Hypergames." *OMEGA* 5(6): 1977. 749–751.

Bryant, J. "All the World's a Stage: Using Drama Theory to Resolve Confrontations." *OR Insight.* 10(4): 1997. 14–21.

Bush, G. Speech by U.S. President in September 1990 following Iraq's invasion of Kuwait. 1990.

von Clausewitz, C. *On War.* Edited with introduction by A. Rapoport. London: Penguin Books. 1968; 1st edition 1832.

Davies, N. *Europe: A History.* London: Pimlico. 1996.

DERA. *Drama-Theoretic Tools for Operational Analysis.* Report under contract number CDA/P/564. 1997.

Fisher, R. & W. Ury. *Getting to Yes.* London: Hutchinson. 1983.

Howard, N. *Paradoxes of Rationality:Theory of Metagames and Political Behavior.* Cambridge, MA: M.I.T. Press. 1971.

Howard, N. "Drama Theory and Its Relation to Game Theory. Part 1:Dramatic Resolution vs. Rational Solution." *Group Decision and Negotiation.* 3: 1994.187–206.

Howard, N. "Drama Theory and Its Relation to Game Theory. Part 2: Formal Model of the Resolution Process." *Group Decision and Negotiation.* 3: 1994a. 207–235.

Howard, N. "Negotiation as Drama: How 'Games' Become Dramatic." *International Negotiation.* 1: 1996. 125–152.

Howard, N. "What Is Drama Theory?" In *Modelling International Conflict: Proceedings of a Conference by the Institute of Mathematics and Its Application.* Southend-on-Sea: IMA. 1997.

Howard, N. "Drama Theory: The Fundamental Theorems" in *Modelling International Conflict: Proceedings of a Conference by the Institute of Mathematics and its Application.* Southend-on-Sea: IMA. 1997a.

Howard, N. "N-Person 'Soft' Games." *Journal of Operational Research Society.* 49: 1998. 144–150.

Howard, N., P. Bennett, J. Bryant, and M. Bradley. "Manifesto for a Theory of Drama and Irrational Choice." *Journal of Operational Research Society.* 44(1): 1992. 99–103.

Joint Chiefs of Staff. *Joint Vision 2010.* 1997.

Kahn, H. *On Escalation: Metaphors and Scenarios.* New York, Washington: Frederick A. Praeger. 1965.

Kuhn, T. S. *The Structure of Scientific Revolutions.* Chicago: University of Chicago Press. 1962.

Maxwell, D. "Adaptive Command and Control: Future Mission Statements in Peace Operations." *Proceedings of Third International Symposium on Command and Control Research and Technology,* National Defense University, Washington, DC. 1997.

Cooperation—or Conflict (Newsletter). Birmingham, UK: Nigel Howard Systems. 1992–1997.

Osborne, M., and A. Rubinstein. *A Course in Game Theory.* Cambridge: MIT Press. 1994.

Owen, D. *Balkan Odyssey.* London: Indigo (Cassell Group). 1996.

Popper, K. *The Logic of Scientific Discovery.* London: Hutchinson. 1959, original German edition 1934.

Rapoport, A. *Fights. Games and Debates.* New York: Harper. 1960.

Rapoport, A. *Strategy and Conscience.* New York: Harper. 1964.

Rapaport, A. Preface to Clausewitz *On War.* London: Penguin Books. 1968.

Rogers, P. "War in the 21st Century: Likely Trends and Their Possible Control." In *Modelling International Conflict: Proceedings of a Conference by Institute of Mathematics and Its Applications.* Southend-on-Sea: Institute of Mathematics and its Applications. 1997.

Schelling, T.C. *The Strategy of Conflict*. Cambridge, MA: Harvard University Press. 1960.

Silber, L., and A. Little. *The Death of Yugoslavia*. London: Penguin Books (BBC Books). 1996.

Tait, A. "Drama Without Tears." *Tutorial Papers, Young OR 10*. Birmingham, UK: Operational Research Society. 1998.

von Neumann, J., and O. Morgenstern. *Theory of Games and Economic Behavior*. Princeton, NJ: University of Princeton Press. 1959; 1st edition 1944.

Wentz, L.K. "C3I for Peace Operations: Lessons from Bosnia." *Proceedings of Third International Symposium on Command and Control Research and Technology*. Washington DC: National Defense University Press. 1997.

ABOUT THE AUTHOR

Nigel Howard is Managing Director and Chairman of ISCO Ltd and Visiting Professor at Sheffield Business School, Sheffield Hallam University, UK. His early work in game theory focused on the practical "dilemmas" created when one tries to take a purely "rational" approach to human interactions. His work has remained practical, while remaining focused on this same problem. His methods were applied, in the late sixties, to the first SALT agreement, nuclear proliferation, and Vietnam; they have since been used on many business and governmental policy problems. In recent years he has been the chief developer of a new approach—"drama theory." He has taught at the LSE and the universities of Pennsylvania, Waterloo, Ottawa, and Aston. From 1987 to 1997 he published a bimonthly research letter, *Cooperation or Conflict*, now an Internet site (www.nhoward.demon.co.uk\drama.htm). From 1997 onwards he has applied a technique derived from drama theory—Confrontation Analysis—to defense problems, in particular to Peace Operations. Currently he is engaged in a project for a Command and Control System for Confrontations. In the non-defense sector, he is involved in work on both mergers and acquisitions, and also in designing a "virtual hospital" for medical personnel training.

Order Form

The publications listed on this form are available at no cost through CCRP. Simply mark the publication(s) you would like to receive and return these sheets to the following address:

CCRP Publications Distribution Center
c/o Evidence Based Research, Inc.
1595 Spring Hill Road, Suite 250
Vienna, VA 22182-2216

or fax to:

(703) 821-7742

❏   Please change my listing in your database.
❏   Please add my name to your database.
❏   Please remove my name from your mailing list.

Name _____

Title _____

Company _____

Address _____

City _____ State _____ Zip _____ Country ____

Phone _____

E-mail _____

Area(s) of Expertise _____

I would like to see more publications on the following subject(s):

_____

I use these publications primarily for the following reason(s):
❏   Personal reference
❏   Work reference
❏   Course textbook

For further information on CCRP, please visit our website at
www.dodccrp.org

Command and Control

- ❏ Command, Control, and the Common Defense (Allard)
- ❏ Command and Control in Peace Operations Workshop
- ❏ Command Arrangements for Peace Operations (Alberts & Hayes)*
- ❏ Complexity, Global Politics, and National Security (Alberts & Czerwinski, eds.)
- ❏ Coping with the Bounds (Czerwinski)

Information Technologies and Information Warfare

- ❏ Behind the Wizard's Curtain: An Integration Environment for a System of Systems (Krygiel)
- ❏ Defending Cyberspace and Other Metaphors (Libicki)* *limited supply*
- ❏ Defensive Information Warfare (Alberts)*
- ❏ Dominant Battlespace Knowledge (Johnson & Libicki)*
- ❏ The Information Age: An Anthology on Its Impacts and Consequences (Papp & Alberts, eds.)*
- ❏ Information Warfare and International Law (Greenberg, Goodman, & Soo Hoo)
- ❏ The Mesh and the Net: Speculations on Armed Conflict in a Time of Free Silicon (Libicki)
- ❏ Network Centric Warfare: Developing and Leveraging Information Superiority (Alberts, Garstka, & Stein)
- ❏ Standards: The Rough Road to the Common Byte (Libicki)*
- ❏ The Unintended Consequences of Information Age Technologies (Alberts)*
- ❏ What is Information Warfare? (Libicki)*

Operations Other Than War

- ❏ Confrontation Analysis: How to Win Operations Other Than War (Howard)
- ❏ Doing Windows: Non-Traditional Military Responses to Complex Emergencies (B. Hayes & Sands)
- ❏ Humanitarian Assistance and Disaster Relief in the Next Century (Sovereign)
- ❏ Interagency and Political-Military Dimensions of Peace Operations: Haiti— A Case Study (Daly Hayes & Wheatley, eds.)
- ❏ Lessons from Bosnia: The IFOR Experience (Wentz, ed.)
- ❏ NGOs and the Military in the Interagency Process (Davidson, Daly Hayes, & Landon) *limited supply*
- ❏ Operations Other Than War* *limited supply*
- ❏ Shock and Awe: Achieving Rapid Dominance (Ullman & Wade)* *limited supply*
- ❏ Target Bosnia: Integrating Information Activities in Peace Operations (Siegel)

Proceedings of the Command and Control Research and Technology Symposium (CD-ROM Only)

- ❏ 1998/4th International
- ❏ 1999

*Published in conjunction with the NDU Press.